August Henry Keane

Eastern Geography

A Geography of the Malay Peninsula, Indo-China, the Eastern Archipelago, the

Philippines and New Guinea

August Henry Keane

Eastern Geography
A Geography of the Malay Peninsula, Indo-China, the Eastern Archipelago, the Philippines and New Guinea

ISBN/EAN: 9783743320079

Manufactured in Europe, USA, Canada, Australia, Japa

Cover: Foto ©berggeist007 / pixelio.de

Manufactured and distributed by brebook publishing software
(www.brebook.com)

August Henry Keane

Eastern Geography

EASTERN GEOGRAPHY.

THE EASTERN ARCHIPELAGO.

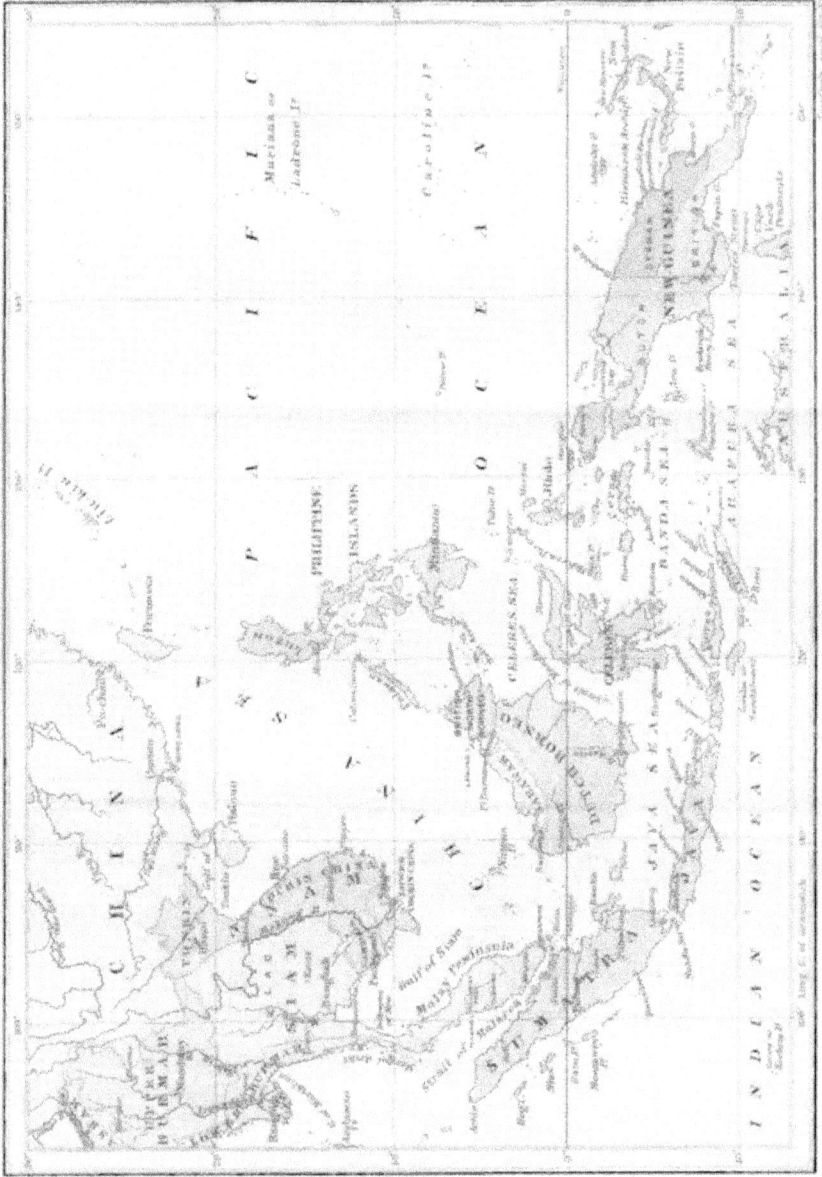

London : Edward Stanford, 55, Charing Cross, S.W.

EASTERN GEOGRAPHY.

A GEOGRAPHY

OF

THE MALAY PENINSULA, INDO-CHINA,

THE EASTERN ARCHIPELAGO, THE PHILIPPINES,

AND NEW GUINEA.

BY

Prof. A. H. KEANE, B.A., F.R.G.S.,

VICE-PRESIDENT OF THE ANTHROPOLOGICAL INSTITUTE ;
AUTHOR OF
"ASIA," IN STANFORD'S COMPENDIUM SERIES.

WITH A MAP.

LONDON :
EDWARD STANFORD, 55, CHARING CROSS, S.W.
1887.

PREFACE.

This work, it is hoped, may be only the first **of a series on** EASTERN GEOGRAPHY, **the** idea, and to some **extent the plan, of** which are due **to the** enlightened public spirit **of the Hon. A.** M. Skinner, President (1885) of the Straits **Branch of the** Royal Asiatic Society. During his official connection with the Administration of Singapore, that gentleman has practically co-operated in supplying a want which **the** Governor, Sir F. Weld, **had** long desired to supply. Under these auspices there **appeared in** 1884 at Singapore a treatise on the Malay Peninsula which **has** served as the groundwork of the first part of the present volume. Some materials collected on the spot were also kindly placed at my disposal, of which **I** have gladly availed myself in the treatment of other sections.

For the general plan and composition **of the** volume in its present form I must in other respects accept **the** entire responsibility. My primary aim has been to produce **a** work which may **meet** the requirements of teacher and pupil **in** the Straits Settlements, and in the other colonies directly interested in the regions here dealt with. At the same time, these regions, notwithstanding their growing political and commercial importance, continue **to be** handled in such a perfunctory manner in popular works **at** home, that English students may also perhaps be **glad** to welcome a work which can at least claim to be the first exclusively devoted to those remote lands.

A glance at the Contents, which by a process of double pagination have been so disposed as to dispense with an index, will at once reveal the general arrangement of the subject matter. Here clearness and uniformity have been the main considerations, while in the treatment especially of the physical and biological sections an attempt has been made to break away from the crude methods still lingering in our schools, and to bring the matter more into harmony with the views of the Ritters, Peschels, Reclus and the other illustrious exponents of the true scientific method. Thus the present conditions are, as far as possible, treated in the light of the past, so that a relation between cause and effect takes the place of a bald statement of facts. In this way the slow decay of the marvellous Cambojan culture becomes intimately associated with the slow subsidence of the waters, or the upheaval of the land, which converted a former marine inlet into a mere fishing pond (p. 102-3). So also the local phenomenon of the "Sumatras" is brought into direct connection with the climatic, and these again with the geological conditions of North Sumatra (p. 141), and so on.

Another feature is the reference to recent explorers (Forbes, Chalmers, Guillemard, Gill, &c.), in those still little known regions, and even occasional short quotations from their writings. This inspires the teacher with confidence in his guide, and perhaps helps to awaken the interest of his pupil. All the information is as recent and correct as possible, and for that reason the book appeals to many besides pupils and teachers, for whom it was originally intended.

The division of the Eastern Archipelago into three instead of two natural regions (an Asiatic, Oceanic, and Australian) may possibly challenge criticism. But if it teaches teacher and student to think, its purpose will be served, even though the theory itself be rejected.

The orthography was of course a troublesome question, the solution of which has been sought in an eclectic system

leaning towards the suggestions lately published by the Royal
Geographical Society. The indefinite Malay vowel (ĕ) will be
found generally marked in important names (Kĕdah, Sĕnggóra, &c.), but its consistent adoption throughout would have
needlessly overcrowded the pages of the Malay section with
unsightly diacritical marks. For some useful information on
this and other points I have to thank Mr. D. F. A. Hervey of
Malacca, though unable to adopt all his suggestions. The Rev.
James Chalmers has also kindly looked over the proof sheets of
the section on New Guinea.

A. H. K.

UNIVERSITY COLLEGE, LONDON,
March 1887.

ERRATA.

Page 2, line 18 from top, for in Johor read in Moar.

" 39 " 44 " for Abubákar, K.C.S.I. read the Máharája (since 1885 called Sultan) Abubákar, G.C.M.G., K.C.S.I.

" 40 " 20 " for the restoration of Moar to Johor read the provisional and temporary administration of Moar by the ruler of Johor.

" 51 " 33 " for north-east read south-east.

" 54 " 40 " " Hung-hao " Hung-hoa.

" 89 " 42 " " 13° 18′ N. lat. " 13° 45′ N. lat.

" 127 " 14 " " west " east.

" 133 " 9 " " Amuntal " Amuntai.

" 135 " 20 " " Kumanis " Kimânis.

" 148 " 18 " " Palan " Pulau.

" 174 " 6 " " 120° E. " 128° E.

CONTENTS.

PART II. INDO-CHINA.

CHAPTER I.

General Survey—Physical Features—Mountain and River Systems—Seaboard—Islands 49

CHAPTER II.

Climate—Flora—Fauna 55

CHAPTER III.

Inhabitants—Burmese—Talaings—Siamese—Annamese —Cambojans 57

CHAPTER IV.

Political Divisions—Burmah—Siam—Annam—Camboja 66

PART III. THE EASTERN ARCHIPELAGO.

CHAPTER I.

General Survey—Distribution of Land and **Water**—
Main Insular Groups—Volcanic Formations—
Geology—Extent—Population 108

General Survey, p. 108; Volcanic Formations, p. 110; Geology,
p. 111; Extent, Population, p. 112.

CHAPTER II.

Climate—Flora—Fauna 113

CHAPTER III.

Inhabitants—Malays—Indonesians—Negritoes—
Papuans 116

The Malays, p. 117; Malayan Groups, p. 118; the Indonesians,
p. 119; Negritoes, p. 120; Papuans, p. 121.

CHAPTER IV.

Geographical and Political Divisions—Asiatic, Australian, and Oceanic Natural Divisions—Dutch, Spanish,
English, German, and Portuguese Territories 124

1. ASIATIC DIVISION The large Sunda Group with Bali and
Islands adjacent to Sumatra—the Philippine and Sulu Archipelagoes, p. 125; BORNEO, p. 126; Rivers, p. 129; Lakes, p.
129; Climate, p. 130; Flora and Fauna, p. 130; Inhabitants,
p. 131; Dutch Possessions, p. 132; Banjer-Masin, p. 133;
Kútei, p. 133; British Settlements in Borneo, p. 134; Saráwak, p. 134; British North Borneo, p. 135; Labuan, p. 136;

PART I.

THE MALAY PENINSULA.

CHAPTER I.

GENERAL SURVEY—PHYSICAL FEATURES—MOUNTAIN AND RIVER SYSTEMS—SEABOARD—ISLANDS—THE ISTHMUS OF KRA.

Position—Extent.—The Malay Peninsula, the *Tânah Malâyu*, or "Malay Land" of the natives, forms the southernmost extension of the great peninsular region of Indo-China, with which it is connected by the Isthmus of Kra (Kraw). At the narrowest point of this isthmus the river Pakshan marks the natural and political boundary towards British Burma on the west side; but on the east the frontier towards Siam is indicated by no physical or conventional line. South of Kra the Peninsula projects for about 600 miles first south, then south-east nearly parallel with Sumatra, terminating at Cape Tanjong Bûlus in 1° 16' 12" N. latitude. Here is the southernmost extremity of the Asiatic continent, which, however, is geologically continued to the island of Billiton (Bilitong), and includes the neighbouring archipelagoes of Bentan, Lingga, and Banka, all now severed from the mainland. The Peninsula, which is washed by the Bay of Bengal and Strait of Malacca on the west, by the Gulf of Siam and China Sea on the east, gradually widens from about 40 miles at Kra to about 200 miles between the Dindings and Tringgânu, again contracting further south to a mean breadth of under 100 miles in Johor. The total area is somewhat over 75,000 square miles, with an estimated population of at least 1,200,000, or about 15 inhabitants to the square mile.

Mountain Systems.—Malay land forms geologically a southern extension of the mountain system, which separates the Salwin and Menam river basins. It consists mainly of continuous ranges running

B

in a line with the continental axis and forming a distinct water-parting between the streams flowing east and west to the surrounding seas. The western range continues unbroken from the interior of Kedah (6° N.) to the interior of Malacca (2° N.), reappearing at intervals further south in Johor and even in the insular peaks beyond. The central upland region is skirted on either side by low-lying coastlands of varying breadth and of recent formation, which alone are cultivated and inhabited by settled populations.

The height of the main central range increases towards the wider **parts of** the Peninsula, culminating in Kedah and Perak, where several peaks are known to range from 5000 to 8000 feet and upwards. The principal summits, some of which have been ascended in recent years, are Mount Robinson or Riam (about 8000 feet) in south Perak ; Titi Wangsa (6840) between Kedah and Perak ; Ulu Temeling (6435) and Bubo (5650) near the right and left banks of the Perak river respectively ; the Slim range (6000 to 7000) in south-east Perak . Chimberas (5650) in Selangor ; Berembun (about 4000) **in** Sungei Ujong ; Lĕdang, or Ophir (4200) in Johor, until recently supposed to be the highest point in the Peninsula ; Blumut (3200) in south Johor, at the source of the river Johor.

East of the central range, and many miles inland from Perak, **on** the east side of the river Pahang, near the west frontier of Tringgánu and Kelantan, stretches the still unexplored Tahan chain, which was described **in** 1875 by the traveller Miklukho-Maclay as the loftiest crest in the whole Peninsula. Recent information tends to **confirm this** view, and it now seems probable that the highest of the **peaks** exceeds 10,000 feet.

Apart from the low-lying coastlands, which vary in breadth from 10 to 25 miles, the whole peninsula is broken and hilly, and everywhere covered with dense forests. The formation is mainly granitic, traversed by veins of stanniferous quartz, and overlaid by sandstone, unfossilised clay slates, laterite or ironstone, and in a few places, principally towards the north, by limestone. Although no trace has been found of recent volcanic action, there are several isolated and unstratified limestone masses from 500 to 2000 feet high of a highly crystallised character with no fossils of any kind.

Mineral Wealth.—The most remarkable geological feature **is** the prevalence of tin, in some **places** associated with gold and galena. The tin occurs throughout **the** Peninsula, reaching as far north as Tavoi (14° N.) in British Burma, and as far south as the Carimons (Kerimun) and Lingga on the equator, and after a break of about 140 miles reappearing in Banka and Billiton islands (3° S.). Where **it has** been observed *in situ*, the bed of the ore, which occurs nowhere else in the Eastern Archipelago, is the **quartz,** which is found penetrating the granite at every elevation. The whole country has been described as " a vast magazine" of this metal, and is now admitted to be the most extensive tin-producing region in the world. But mining operations have hitherto been confined to the deposits near the foot of the hills, in the alluvial ground formed by the decomposition of the encasing rocks. Mines are worked at present in about twenty different localities on both sides, and throughout the length of the Peninsula. The most productive are those of the Siamese provinces in the north-west, Intan, Selâma, Lârut, Kinta, Kwâla-Lumpor, Sungei, Ujong, Pahang, Kelantan, and Patâni.

Gold occurs in several of these districts, **but** especially in Chendras, Taong (near Mount Ophir), Kelantan, and Jelei in the interior of Pahang, the produce of the last-mentioned place commanding a higher price by 5 per cent. than the best Australian gold. Rich galena ore occurs in Patâni. Silver also, the presence of which had been doubted, although the Perak river is named from the Malay word perak, "silver," has recently been found in Lârut associated with the tin ores of that district. Rich galena ore occurs in Patâni, while iron is more abundant even than tin, especially in the southern provinces. Coal is stated to have been recently found to the south of Kra, in Perak, and a few other places. But neither coal nor iron has hitherto been worked in any part of the Peninsula.

River Systems.—Owing to the formation of the land and the somewhat central disposition of the main water-parting, the rivers although numerous are necessarily of short length, and as their mouths are generally obstructed by bars and coral reefs, they are on the whole more useful for irrigation than as highways of communication. Nevertheless some are navigable by light craft for considerable distances, and small steamers have ascended the Bernam between Perak and Selângor for a distance of about 80 miles from the coast. But by far the largest river basins are the Perak on the west and the Pahang on the east slope, each of which comprises an area of drainage over 5000 square miles in extent. The Perak with its chief tributaries, the Plus, Kinta, and Batang **Padang, presents** a total navigable waterway of perhaps 2000 miles.

The **other** chief streams, following the coast from north to south, are the Pakshan on the northern frontier ; the Mûda, flowing between Kedah and the province of Wellesley ; the Krian and Lârut in Perak ; the Selângor, Klang, Langat, Linggi, and Moar, all on the west coast ; the Johor, whose estuary faces Singapore ; the Patâni, the Kelantan, with its large tributary the Lebih, the Kemâman, Cherâting, Rumpen, and Endau, **all on** the east coast.

Most of these rivers have their course, not east and west, but more or less synclinal with the mountain-ranges from north-east to south-west on the west side, and from south-west to north-east on the opposite side of the Peninsula. A consequence of this disposition of the river basins is, that at some of the principal points of the system the streams flowing from the same water-parting north to the China Sea and south to the Bay of Bengal have their upper waters almost contiguous. Such is the case, for instance, with the rivers Pahang and Slim in 5° North, and the rivers Serting and Moar in 3° North.

Seaboard—Islands.—The coast on both sides, but particularly on the west, is almost invariably marshy and alluvial. The flat, unbroken seaboard, scarcely raised above sea level, is generally overgrown with mangroves for some four or five miles inland. In some parts these low-lying plains expand to a breadth of **25** or 30 miles, but they are usually much more contracted. On the east coast the hills approach at several points close to the shore, a disposition partly due perhaps to the influence of the north-east monsoon. Here the chief headlands are Capes Carnom, Patâni, Tringano, and Romania, to which correspond on the opposite side the promontories of Sâlang, Kalang, Rachado, and Polus (Bûlus).

In the extreme North both sides of the Peninsula are fringed by clusters of innumerable reefs and islets lying close to the shore. Further seawards is a second barrier of larger islands in the Gulf of Siam, of which the chief are Taw, Carnam (Samui), and Quin. In the Bay of Bengal there also runs a second chain, forming a southern extension of the Mergui Archipelago. But beyond this insular region the coast is generally free from islands, except at the southern extremity of the Peninsula, where are clustered the Singapore, Bintang (Bentan), Bûlang, and Carimons (Kerimon) groups. Elsewhere the largest islands are Junk Ceylon (Ujong Sâlang), Lengkâwi, and Penang (Pinang) on the west side; Tantalam, the **Great and Little** Redangs, Tioman, and Tinggi on the **east side.** Their geological formation and general disposition parallel with the seaboard show that all these groups are mere fragments of the mainland, with which some of the largest, such as Sâlang, Singapore, and Tantalam, are almost contiguous. The Strait of Singapore presents the aspect rather of a river than of a marine channel, running for over 30 miles transversely with the main peninsular axis, with a mean breadth of little over 1500 yards.

Isthmus of Kra.—These islands thus bear somewhat the same relation to the whole Peninsula that this region will present to the Asiatic mainland whenever the projected canalisation of the Isthmus of Kra is effected. By a ship canal at this point the voyage from Calcutta to China would be shortened by 660 miles, and that between Burma and Bangkok by 1300 miles. The original scheme, proposed by Tremenhere was to dredge the river Pakshan as far as the village of Kra, and then tunnel the highest point, thus reaching the Gulf of Siam by the alluvial plain of the Chumpong river. Others suggested a point further south, where the Pakshan is everywhere at least 30 feet deep; while the French engineers Deloncle and Orn prefer an intermediate route from the Pakshan below the rapids to Tasan on the Tayung, or Upper Chumpong. But none of the schemes hitherto proposed have been found quite practicable, and since the more definite surveys of 1882-3, the project has been abandoned.

CHAPTER II.

CLIMATE—FLORA—FAUNA.

Climate.—The climate is everywhere moist and hot, though seldom malarious, even along the low muddy banks near the **coast.** Nor is the heat so intense as in South Arabia and other regions much further removed from the equator, the mean annual temperature even on the lowland plains not exceeding 80° F. There is, strictly speaking, no winter, nor even any very distinctly marked rainy season, the alternate north-east and south-west monsoons distributing the moisture over the east and west slopes throughout most of the year.

The average number of rainy days is about 190, giving for the whole Peninsula a mean rainfall of from 90 to 130 inches. The **west** coast is generally well sheltered, although exposed to sudden **squalls of** short duration, known as "Sumatras," from the direction **whence** they blow. On the other hand, the east coast is entirely **closed** to navigation for about five months, during the prevalence of the north-east monsoon sweeping over the Gulf of Siam and China Sea.

Flora.—Except in some limestone tracts, especially in Perak and Kedah, the soil is not very rich. But although not at present yielding sufficient rice for the local demand, the Peninsula appears to be capable of **growing** almost every tropical plant. The land is almost everywhere clothed with a magnificent tropical vegetation, in which the most characteristic and useful growths are several varieties of gutta-percha (*getah*, here first discovered), the camphor tree, ebony, eaglewood, sapan, ratan, nibung, bamboo, nipa-palm, cocoa-nut, areca, and gomuti. The nutmeg, cinnamon, and clove have been introduced, and thrive well, although the nutmeg is subject to a leaf disease. Indigo, gambier, pepper, the sugar-cane, tea, coffee, and tapioca have also been acclimatised. A species of climbing indigo and the wild nutmeg are indigenous, as are also the characteristic durian and mangosteen fruit trees. The most generally cultivated plants are rice, sugar-cane, cotton, tobacco, yams, batata, cocoa and areca palms. With the spread of agriculture and mining operations the primeval forests tend to disappear, and in many districts extensive tracts have already been cleared by the Chinese miners, who recklessly cut down the finest trees to serve as fuel for smelting the tin ores.

Fauna.—The Fauna of the Peninsula, which is unusually rich, is allied, like the flora and the inhabitants, partly to that of the Eastern Archipelago, partly to the Asiatic mainland. Here are the one-horned rhinoceros, Malay tapir (teno), elephant, and hog, all of the same species as those of Sumatra. Here are also a small bear (bruang), found elsewhere only in Borneo, and the Sunda ox of Java, besides two kinds of bison, said to be peculiar to the Peninsula. On the other hand, the Asiatic tiger has extended his range throughout the whole region, even crossing over to Singapore and other adjacent islands. Of quadrumana there are no less than nine species, including the kukang (*Lemur tardigradus*), a so-called chimpanzee (*Simia troglodytes*), the black and white ungka, but apparently not the orang-hútan, although the term is in common use, and applied by the Malays in its natural sense of "wild men" to the wild tribes of the interior.

Of birds perhaps the most characteristic are the rhinoceros hornbill (*Buceros*), the bangau or Javanese stork, the argus and pencilled pheasants, some birds of paradise (*Paradisea regia* and *P. gularis*), the myna or grackle (*Gracula religiosa*), the murei or dial bird (*Gracula saularis*), besides kingfishers, fly-catchers, doves, and pigeons in endless variety. The islands are frequented by the *Hirundo esculenta*, or swallow that builds edible nests, and the forests swarm with coleoptera, lepidoptera, and other insects, including the magnificent butterfly, *Ornithoptera Brookeana*, till recently supposed to be peculiar to Borneo. The surrounding waters are inhabited by the halicore, or "mermaid," a sirenian, whose Malay name of *duyong* has been corrupted to *dugong* in English.

CHAPTER III.

The Negritoes.—Apart from the Chinese, Klings (Indians), and other recent settlers, the inhabitants of the Peninsula belong to three distinct ethnical stocks—the NEGRITO, TAI or SIAMESE, and MALAY. The Negritoes, now reduced to a few fragmentary groups scattered over the more inaccessible parts of the interior, represent the true aboriginal element, and appear to belong to the same primitive type as the so-called "Mincopies" of the Andaman Islands, and the Aetas of the Philippine Archipelago. Their presence in the peninsula, long doubted by ethnologists, has been fully confirmed by the researches of Miklukho-Maclay, and other recent explorers. North of the Pérak they are known by the name of *Semang* (*Samang*), south of that river by that of *Sâkei*, and south of Malacca as *Orang Benûa*, or "Men of the Soil." At the same time, these and other terms, such as the local Besisî and the more comprehensive Mentra, are applied by the civilised Malays somewhat vaguely to all the wild tribes of the interior, whether of Negrito or Malay origin. Nor is this surprising, seeing that the two races themselves, who have been in contact for ages, have become largely intermingled and assimilated in appearance, customs, and even in speech. "Purely anthropological observations," remarks Miklukho-Maclay, "lead me to accept the supposition of a Melanesian [Negrito] element, a remnant of the original race, which through intermixture with the Malays is being more and more supplanted. In the mountains of Pahang and Kelantan, as far as Senggôra and Ligor, I have discovered a Melanesian population. This people undoubtedly belongs to the Melanesian stock " (*Ethnological Excursion in Johor*). All the Negrito tribes are in an extreme low state of culture, holding aloof from the settled populations, living entirely on the chase, and pursuing the game with poisoned arrows.

The Siamese and Sam-Sams.—Excluding the Negrito element, insignificant numerically, and without social or political influence of any kind, the whole of the Peninsula is occupied by the Siamese and Malay races. The former, intruders from Siam in comparatively recent times, hold the northern division as far south as the borders of Kedah and Patáni, or about 7° N. latitude. The latter, also doubtless intruders from the North in remote prehistoric times, prevail throughout the southern and much larger division, to which alone the term "Malay Land" is strictly applicable. The transition between the two races is effected by the Sam-Sams, a half-caste Malayo-Siamese people, lying mainly between the 7th and 8th parallels. These Sam-Sams appear to be mostly Buddhists like the Siamese, whom they also resemble in their customs, traditions, and national aspirations. In speech also they are at least as much Siamese as Malay, both languages being equally current amongst them. The pure Siamese themselves differ in no material respect from the rest of the inhabitants of Siam, and need not here be **further** considered.

THE MALAYS.

The Malays (Orang Maláyu, "Malay men") **are** the dominant people, not only in the southern section of the Peninsula, but throughout **the** Eastern Archipelago, where they are diversely inter-mingled with other races, and where they have represented the local **cultured** element for over two thousand years. The Malays proper, that **is,** those who call themselves by this name, who speak the standard Malay language, and who possess a common sentiment of national unity, are found in compact masses chiefly in the Malay Peninsula, in the adjacent islands of Pinang, Bintang, Lingga, Biliton, Bangka, and in Sumatra, of which they occupy about one half, mainly in the south, along the east coast, and on parts of the west coast. In these lands alone they **are** really indigenous, **and** regard themselves as the aboriginal population. Elsewhere they are met in scattered communities, chiefly round the coast of Borneo, in the Sulu Archipelago, in Tidor, Ternate, and some other members of the Molucca group, where they are held to be intruders, or immigrants from Sumatra.

Long considered as an independent division **of** mankind, the **Malays** are now more generally affiliated to the Mongol stock, of **which A. R.** Wallace, De Quatrefages, and other eminent naturalists regard **them** as a simple variety more or less modified by mixture with other elements. In fact, the typical Malay can scarcely be distinguished anthropologically from the typical Mongolian. He is described by competent observers as of low stature, averaging little **over** 5 feet, of olive-yellow complexion, inclining to light brown **or** **cinnamon,** brachycephalous or round-headed, with somewhat flat **features,** prominent cheek-bones, black and slightly oblique eyes, small but not flat nose, dilated nostrils, hands and feet small and delicate, legs thin and weak, coarse black hair always lank and round in section, scant or no beard.

The departure from this description so frequently noticed in the Archipelago must be attributed to intermixture with the black Papuan stock in the east, and with a distinct pre-Malay Caucasic element in the west. The presence of this "Indonesian" element, as it is called by Logan and Hamy, may now be regarded as an ascertained fact, the recognition of which will help to remove many of the difficulties connected with the various relations of the Malays to the surrounding races. It at once explains, for instance, the apparent discrepancy between the foregoing description of the ordinary Malay and that of the Battas, Orang Kubu, and many other Sumatran and Bornean peoples described as tall and robust, with regular features, symmetrical figure, light complexion, brown and wavy hair, and general European appearance.

These considerations also enable us to fix the true centre of dispersion of the Malay race rather on the mainland than in Sumatra, contrary to the generally received opinion. If they are physically allied to the Mongol stock, it is obvious that the earliest migration must have been from high Asia southwards to the peninsula, and thence to Sumatra, possibly at a time when the island still formed part of the continent. The national traditions of a dispersion from Menangkabau or Palembang in south Sumatra must accordingly be understood to refer to late movements, and more especially to the diffusion of the civilised Malay people, who first acquired a really national development in Sumatra in comparatively recent times. From this point they spread to the Peninsula, to Borneo, Sulu, and other parts of Malaysia apparently since their conversion to Islam, although other waves of migration must have reached Further India, if not from the same region at all events from Java, at much earlier dates. The impulse to these earlier movements was due to the introduction of Indian culture through the Brahman and Buddhist missionaries perhaps two or three centuries before the Christian era.

During still more remote pre-historic times various sections of the Malay and Indonesian stocks were diffused westwards to Madagascar, and eastwards to the Philippines, Formosa, Micronesia, and Polynesia. This astonishing expansion of the Malaysian people throughout the oceanic area is sufficiently attested by the diffusion of a common oceanic (Malayo-Polynesian) speech from Madagascar to Easter Island, and from Hawaii to New Zealand.

The Malays proper have long been divided into three distinct social groups:—The *Orang Benûa*, or "Men of the Soil," that is, the uncivilised wild tribes of the peninsula; the *Orang-laut*, or "Men of the Sea," that is, the semi-civilised floating population; and the *Orang Maláyu*, or Malays in a pre-eminent sense, that is, the civilised Malays with a culture, a literature, and a religion. The Orang Benûa, called also *Orang-Gunung*, or "Highlanders," and even *Orang-Utan*, or "Wild Men," constitute the aboriginal Malay element, which has hitherto remained unaffected by foreign influences, and which is still grouped in small tribes at a very low stage of culture, living mainly by the chase, and almost destitute of social organisation. They are found chiefly in the more inaccessible wooded uplands of the peninsula and Sumatra, in the former region more or less intimately associated for ages with the Negrito tribe, and in the latter island apparently the sole occupiers of the land from the first.

Intermediate between the Orang-Benûa and Orang-Malâyu are the Orang-laut, or "Sea Gipsies" of former English writers, the "Cellates" of the early Portuguese explorers. But they are no longer the "vile people dwelling more on the sea than on the land," and "living by fishing and robbing;" for piracy has been almost entirely suppressed in these waters, and the Orang-laut have risen considerably in the social scale since the spread of English power and influences throughout Malay land and North Borneo.

This remark is equally applicable to the Orang-Malâyu, or civilised Malays, who first under Hindu and afterwards under Arab influences developed a national life and culture, and founded more or less powerful political states in various parts of the archipelago and throughout the peninsula. At one time there was an impression that they were losing ground, and becoming gradually displaced by the Chinese immigrants into Malaysia. But statistics have shown that this view was groundless, and during the present century the whole Malay race has everywhere displayed an unexpected vigour and vitality. The native populations of Java, Sumatra, and the Peninsula, far from showing any tendency to dwindle away before the Chinese intruders, have multiplied considerably, and are at present probably four times more numerous than at the beginning of the century. In the British possessions of Pinang and Malacca the Malay element has increased from 30,000 in 1800 to nearly six times as many at the census of 1881.

In their temperament no less than in their physical features the Malays still betray their Asiatic origin. They are described as of a taciturn, undemonstrative disposition, little given to outward manifestations of joy or sorrow, yet extremely courteous towards each other, and as a rule kind to their women and children. Slow and deliberate of speech, neither elated by good nor depressed by bad fortune, normally impressive and indolent, they are nevertheless capable of the greatest excesses when their passions are roused. Under the influence of religious excitement, losses at gambling, jealousy, or other domestic troubles, they are at times seized by the so-called "amok" fever, when they will rush wildly through the crowded bazaar armed with their sharp krisses, cutting down all who cross their path with incredible fury and without the least discrimination.

The Orang-Benûa are still nature-worshippers; but the civilised

Malays, together with some of the northern Sam-Sams, universally profess the Mohammedan religion. Until about the year 1260 they were pagans, or followed some corrupt form of Hindu idolatry. But the powerful Sultan Mahmud Shah, having adopted Islam in the 13th century, spread the new doctrine throughout his dominions during his long reign of 57 years. His rule extended over the provinces of Malacca, Johor, Patáni, Kedah, and Pérak on the mainland, the neighbouring islands of Lingga and Bentan, and apparently several districts in Sumatra. The Mussulman faith was thus rapidly diffused throughout the Peninsula, and at the beginning of the 16th century the Portuguese found all the Straits Malays zealous followers of the Prophet, while a large portion of southeast Malaysia was still pagan.

Apparently to the Malay stock must be affiliated the primitive community of troglodytes, who occupy the ten small islands in the inland sea of Talé-sab, recently, for the first time, explored by Mr. Davidson and MM. Deloncle and Macey. The archipelago everywhere abounds in caves, in which the natives are born, live, and die, occupied exclusively in collecting and preparing for the Chinese market the edible swallows' nests covering the walls of their rocky dwellings. In gathering the nests, from which a revenue of nearly £30,000 is derived, they display extraordinary agility and hardihood. At one time they seem to have been brought under Hindu influences, for in one of the caves there is a shrine furnished with Brahman religious emblems, and containing two rudely-carved wooden images of great antiquity, representing the king and the queen of the swallows.

The Malay Language.—The Malay language is the most important of the many dialects composing the Malayan section of the Malayo-Polynesian or oceanic family. The area over which it is spoken comprises the peninsula with the adjacent Rio-Lingga Archipelago and other islands, the greater part of the coast districts of Sumatra and Borneo, the Moluccas, the seaports of Java, and to a less extent those of Celebes, besides Tidor, Ternate, and parts of Jilolo. It had already become the general medium of communication throughout Malaysia from Sumatra to the Philippine Islands when the Portuguese first appeared in that region nearly 400 years ago. But before that time there appears to have been no written standard; nor have any monumental records been found with inscriptions written in Malay before the adoption of the Arabic character.

It is not a little remarkable that Malay should have remained unwritten, while the Javanese, the Rejangs and Battas of Sumatra, the Bugis of Celebes, and even the Tagalas of the Philippines all possessed peculiar characters, all doubtless of Hindu origin. But with the Mohammedan conversion the Perso-Arabic alphabet was introduced amongst all the civilised Malays. Malay is essentially a dissyllabic language, harmonious, and of simple structure. From the Hindus, who appear to have settled in Sumatra, Java, and Bali about the 3rd century, if not earlier, the native dialects adopted a large number of Sanskrit terms ; and since the 13th century many Arabic words and expressions have, under the religious influence of Islàm, found their way especially into the literary Malay language. No real distinction can be drawn, as is sometimes done, between *High Malay* and *Low Malay*, as between Kawi and modern Javanese. Low Malay is not a distinct dialect, but merely a colloquial form serving as a medium of intercourse between the natives and Europeans.

Malay is said to be spoken with the greatest elegance in the Rio-Lingga Archipelago, and among the Malay States along the south-west coast of the Peninsula.

The Chinese.—Besides the indigenous ethnical groups of Negritos, Malays, and Siamese, the Peninsula is inhabited by a large number of immigrants from the surrounding regions and from Europe. By far the most numerous of these foreign peoples are the Chinese, who are already in an absolute majority in some districts, and who form about one-third of the whole population. Many of them have married Malay women, and their offspring while remaining Chinese generally adopt the local speech and usages. With the characteristic versatility of the race, they adapt themselves to all conditions of life, and are specially noted for their skill and enterprise as miners, artisans, traders, and agriculturists. If treated with justice and kindness, they are on the whole a peaceful and law-abiding people; but troubles have occasionally arisen, especially through the influence of their secret societies, the members of which often bind themselves to recognise no civil jurisdiction except the authority of these associations. Peace, however, has hitherto been maintained, partly by the policy of dividing these dangerous elements into hostile groups, partly by securing the co-operation of the wealthy Chinese capitalists and traders, who are admitted under various titles into the British administration.

The Klings, &c.—Next in importance to the Chinese are the Hindus, or natives of British India, who are also divided into distinct groups according to the countries whence they have immigrated. The term *Kling*, a contracted form of *Kalinga*, or *Telinga*, that is, "Telugu," is commonly applied to all the Dravidian communities of Telugu and Tamil speech, those speaking Hindustáni

being called " Bengali," the Gujuráti " Orang Bombei," and the
Cingalese " Orang Selon." Amongst the Indians must be included
some Santhals, Kols, and other low caste tribes employed on the
plantations.

Other foreign elements are the Arabs, generally intermingled with their
Malay co-religionists ; the Jews and Armenians, mostly traders who keep
aloof from the surrounding populations ; the so-called " Portuguese " of
the Straits Settlements, who have become darker and often far more
degraded than the Malays themselves, while claiming descent from the
Albuquerques, Castros, Souzas, and other famous pioneers of European
culture in the far East : lastly the English, chiefly merchants, officials,
and planters, but nowhere forming permanent local communities.

CHAPTER IV.

POLITICAL DIVISIONS—SIAMESE AND BRITISH DIVISIONS—RESOURCES—
TRADE—GOVERNMENT—HISTORY.

POLITICALLY the Peninsula is partly held directly by Great Britain
and Siam, and partly divided amongst a number of petty Malay
States, either tributary to or in treaty with these paramount
powers.

The influence of Siam extends over the whole of the Northern
Division, although south of the 7th parallel, where the Siamese race
gives place to the Malayan ; this influence is little more than a
nominal and traditional ascendancy, such as a great power must
necessarily exercise over a small neighbouring State, symbolised by
the old custom of presenting a triennial gold flower to the king of
Siam. But even this custom never extended south of a line drawn
from Kedah Peak on the west coast (5° 40' N.) to the northern
frontier of Pahang (4° 35' N.), which defines approximately the
southern limits of all the land more or less tributary to Siam.

The rest of the Peninsula, which alone belongs to the British
political system, is occupied partly by the British possessions,
grouped under the collective name of the " Straits Settlements " ;
partly by Perak and some other protected States on the west coast,
which are now in effect under British administration ; partly by some
more or less independent Malay States, which must also be regarded
as forming part of the general British protectorate over the whole of
the Southern Division of the Peninsula.

The northern or Siamese section is nearly 40,000, the southern or British nearly 35,000 square miles in extent. But although the former has a larger area, it contains only about one-third of the whole population, and a still smaller share of the trade of the Peninsula.

NORTHERN OR SIAMESE DIVISION.

The Siamese have for some centuries been connected with the northern districts of the Peninsula, first apparently as settlers, and subsequently, down to quite recent times, as conquerors. Since the decline of the Pegu power they have claimed the suzerainty over the littoral of the narrow portion of the Peninsula north of 7° N. lat., which is approximately the southern limit of the race. They also exercise a less defined ascendancy over Kedah on the west, and the Malay States on the east side between Senggöra and Pahang (4° 35′ N.).

Some confusion prevails regarding the nomenclature and sub-divisions of the various States and territories comprised in the Northern Division of the Peninsula. But the subjoined table, which is based on high Siamese authority, appears to contain all the administrative provinces, sub-divisions, and more or less independent states forming part of the Siamese political system.

Provinces administered by Siam—

WEST COAST.	EAST COAST.
Kra with Renong	Pran
Takua-pah	Kumnetne-pe-kun
Takua-tung	Prachuep-ke-re-kan
Pangna	Chumpan
Puket (Junk Ceylon)	Langsuan
Trang	Chaiya
Satun (Setul)	Kamchonedit (Bandon)
Perlis	Plean with Patelüng

Siamese and Malay Tributary States—

Ligor (Lakton)
Senggöra (Songkla)
Patâni {
  Chana (Chenai)
  Nong-chik (Nochi)
  Tâni (Patâni)
  Jering
  Reman
  Sai
  Bana
  Jalap (Jalo)
} EAST COAST.
Kedah   WEST COAST.

Guaranteed Malay States—

Kelantan
Tringgánu with Kemaman } EAST COAST.

Subjoined is a brief account of the more important of these provinces and States, many of which are under Siamese or Chinese rulers, who bear the Siamese title of "Phya." The population, which is estimated at about half a million, consists mainly of Siamese, Chinese, and Sam-Sams in the northern, and Malays in the southern districts.

Kra with Renong.—The isthmus, forming the Northern Division of Lower Siam, lies between 12°—9° N. lat., with an average breadth of about 60 miles. Near the centre on the north bank of the river Pakshan is the town of Kra, whence it takes its name. In the neighbourhood coal is said to have recently been discovered.

Renong, one of the chief places on the west coast, is a tin-producing district, inhabited chiefly by Chinese. It is situated on the south bank of the R. Pakshan. Jointly with Trang it forms a feudatory State under a Penang-born Chinaman, who takes the title of Rája.

Puket or Junk Ceylon (*Ujong Sálang*), a large island occupying a conspicuous position in 8° N. lat. at the north-west elbow of the Peninsula, is separated from the mainland by the narrow Papra Strait. It is 40 miles long by 15 broad, and comprises the two sub-divisions of Talang and Tongka, which formerly belonged to the Rája of Kedah, but are now administered by Siam. The chief place is Puket, on the sheltered east side, where the Siamese commissioner usually resides. A large Chinese population is here engaged in tin-mining, the produce of which is brought chiefly to Penang in return for opium and piece-goods. Other exports are edible birds'-nests, beche-de-mer, and elephants' teeth. The strait and harbour of Papra are accessible at spring tides to ships drawing 20 feet of water.

Kedah, the Portuguese *Quedah* and Siamese *Sai*, is the only **Malay State** tributary to Siam on the west coast. It lies between Trang and Perak, stretching for 120 miles between 7°—5° 30′ N. lat., and for 25 to 30 miles inland, with an estimated area of nearly 4000 square miles, or probably 5000, including the adjacent islands of Lengkawi, Trutao, and a few smaller groups. It comprises three provinces named after the rivers Setúl to the north, Perlis in the centre, and Kedah to the south. The land is less mountainous than most parts of the peninsula, the chief eminence being Jerei, or Kedah Peak (4000 feet), and it is watered by 26 rivers, six of which are considerable, but all of them obstructed by bars at their mouths. Between the mainland and the Lengkawi islands there is also an extensive mud bank preventing vessels from approaching **nearer** than four miles **to the coast.**

The old town, called "Quedah" by the Portuguese, was situated in latitude 6° North upon one of the largest rivers, supposed to be the Merbu, which was navigable for vessels of 300 tons burden.

The geological formation of Kedah, generally speaking, is granite, and in places tin is found and, it is believed, gold. But the more important tin mines are just beyond the Kedah frontiers; and this holds true both to the north, the east, and the south. Limestone crops out in a remarkable manner at Gúnong Wang, on the river Giti, a tributary of the Muda, and at Gúnong Geriyang, or Elephant Mount. The vegetable products are the usual ones of the Peninsula, the country being particularly well adapted for growing rice. Fruit trees of all kinds, especially the mangostin and durian, grow to great perfection. Among its wild animals, the elephant is common, and is used as a beast of burden. Cattle and buffaloes are abundant in the domestic state.

The inhabitants consist of Malays; of a few Sam-Sams or mixed Siamese in the north, who are usually Mahomedan here, and speak both Malay and Siamese; of the Peninsula Negritos; of half-caste Telingas (Klings), speaking both Tamil and Malay; and of a few Chinese.

The capital is Kota Star, or Alor Star, on a river of no great size, though one of the largest of the country, to the north of the conspicuous Mount Jerni. It has for some years been connected, by the rough road already mentioned, with Senggora on the east, the nearest Siamese town of importance. This is at present the only road across the peninsula. The river Muda, the frontier of Province Wellesley, is navigable for small boats **to** Baling, distant about 60 miles east. This place is of some importance as the frontier station, near the point where Kedah, Perak, and Patáni meet; and from Baling the Muda river is used to carry to market at Penang the tin which is found in unusual abundance at Klian Intan and Kroh, on the east or Patáni side of the dividing range of Titi Wangsa.

The country at the back of Province Wellesley is also known to be rich **in** tin at Serdang and Kulim, but it has not yet been fully developed.

It follows from the position of Kedah that its trade is almost exclusively with Penang, with which port communication by steamer is now **easy** and frequent. The exports consist principally of tin, rice, bats' manure **(from** the lime-stone caves), and jungle produce.

The history of this State, as of all the others of the Peninsula to the north **of** Malacca, is full of obscurity. Colonel Low discovered in the forests **some** remains of what he supposed to be Buddhist temples, and some inscriptions in the Pali character, indicating not Malay but Siamese culture. It seems probable that even so late as the beginning of the sixteenth century the Malays here had been but partially converted to Mahomedanism. The earliest authentic information we have of Kedah is from the Portuguese writer Barbosa, whose manuscript is dated "Lisbon, 1516," and he describes it as "a place of the kingdom of Siam," and makes mention of a "Seaport called *Quedah*, to which **an** infinite number of ships resort, trading in all kinds of merchandise." Kedah, in common with all the other northern states of the peninsula, has probably been always more or less tributary to Siam, and being, with Patáni, the most northern of all the Malay States, it has been most subject to its direct influence. But the policy of the Siamese

Government here, as elsewhere, has been to leave the extraneous races,
comprised within the dominion it assumes, to the administration of their
own rulers; the Malayan Rája of Kédah is thus an hereditary and quasi-
independent sovereign. In token of his dependence on Siam he has
always sent the King, once in three years, an offering consisting of an
artificial flower of gold. Nothwithstanding this, the Rája alienated to the
Indian Government in 1786 (Captain Light being the Agent) the island of
Pinang, and subsequently, in 1800, Province Wellesley on the mainland,
without reference to Siam, whose alleged suzerainty was neither well
understood nor much enforced at that time. By the cession of Pinang,
Kédah lost some of its trade, and though the Rája seems to have acted
within his rights, he evidently incurred Siam's displeasure. In 1821, the
Siamese from Ligor invaded the country, overran it, and after an occupation of several years, abandoned after ruining it. The prince fled to
Pinang for protection, and there received an asylum. His line was
restored after many years; but the tendency of the Government at Bangkok to interfere in Kédah affairs has since been accentuated, the King of
Siam even claiming to nominate as well as confirm the Rája.

Ligor, the Siamese LAKHON, is the chief Siamese province in the
north-east part of the isthmus. It was founded four centuries ago
by the King of Ayuthia, and nearly three-fourths of the inhabitants
are still Siamese. The capital is Ligor, on the north side of Lakhon
Bight, 8° 17′ N. lat., 100° 12′ E. long. Here resides the Chow Phya,
or governor, who rules almost absolutely, with power of capital
punishment.

Sĕnggóra, the Siamese SONGKLA, forms the most southerly
Siamese province on the east coast. It borders on the Malayan
States of the Peninsula, and through its Chinese governor, the King
of Siam has hitherto exercised occasional interference with his
Malayan tributaries. The capital lies under the shelter of Tentálam,
a large flat island stretching along the coast, from which it is separ-
ated by a deep and narrow channel of sweet water fed by the Telung
(Patelung) river, from the Kao Luang mountains.

Inland from this channel are the small and semi-independent
Sam-Sam States of Patelung and Plean, under a Chinese Rája.

The east coast being a completely lee shore, there is scarcely any
communication between these smaller provinces of Siam and the
capital. Nor is there any overland route from Sĕnggóra to Bangkok; but a road was opened in the opposite direction across the
Peninsula to Kédah in 1871, at the time of the King of Siam's visit
to the Straits.

This northern group of provinces and petty states, comprising
altogether some 15 separate divisions, mainly inhabited by Siamese
and Chinese, has a coast line of about 200 miles on the east, and a
little more on the west side, with a total area of some 17,000 square
miles, and a population vaguely estimated at from 150,000 to 500,000.

c

Patâni.—The country, still commonly known as Patâni, lies between Sĕnggôra and Kĕlantan (7°—6° 20′ N.) with a coast-line of about 50 miles on the east side, and an estimated area of 6000 square miles. But, except as applied to the river of that name, the term Patâni is merely a historical expression, under which are now comprised as many as nine distinct petty Siamese-Malayan States or provinces. These administrative divisions, together with some other districts, are under the general but ill-defined control of Sĕnggôra, the seat of government for the south-eastern section of the Siamese system, so far as **any** administrative suzerainty exists.

Since its invasion **and** subjugation by Siam in 1832, Patâni has been broken up into the four sea-board States or divisions, lying from north to south in the following order :—Nong-chik or Tûjong, Patâni, Jering, Sai ; and the five interior divisions—Tipah, Chenai, Jalo, Rêman, Ligei. Of these, Rêman is, even excluding any part of the Pérak watershed, the most extensive, and Patâni with its seaport, is probably still the most populous. Jalo and Ligei are believed to be the richest in minerals.

The principal physical features of the province are the two considerable rivers—the Patâni and the Tĕlûpin—which rise in the same hills and flow nearly parallel to the sea, through a country for the most part flat, but with isolated cliffs and hills.

The Patâni is a long but shallow river which retains the same name throughout its whole length. Its source is said to be in the mountain Jambul Mérak (peacock's crest) about 5° 35′, from which also the northern tributaries of the Pérak flow ; thence it has a northerly course and falls into the Gulf of Siam in 6° 55′ N. The upper waters of the rivers Patâni and Pérak are a labyrinth of streams forming the head-waters of the river system of this part of the Peninsula ; the river Kĕlantan is also said to take its source in the same region.

The Patâni has an extensive delta, intersected by numerous creeks. *Kwâla Tûjong* to the north is the most important estuary, and is navigable as far as Kwâla Nong-chik (Nochi) where it bifurcates from the Patâni.

The Bay of Patâni is formed by the projection of a narrow strip of land about seven or eight miles in length, which, connected with the mainland to the eastward, bends round to the north-west like a horn, and protects the roadstead, so that vessels can at most seasons ride in safety. The western extremity of this projection is called Cape Patâni. The town and port of Patâni is almost all that is left unchanged of the former important State of that name. It was and still is the chief town of the whole of this country. It is situated about two miles from the river's mouth, on the south-east side ; a fair amount of trade is carried on with Singapore and Bangkok, as also with the neighbouring Siamese and Malayan States. The exports are tin, lead, gutta, salt-fish, tiles, earthenware, and timber. The population of the town consists of Malays, Chinese, and Siamese, and is supposed to be from 3000 to 4000. The Malay race preponderates, the Râja himself being a Malay. The active commercial and shipping business is controlled by a " Captain China."

Rêman.—The largest and perhaps the most important of the provinces at present is Rêman, lying to the south-east of the river and bordering on Pêrak, with which it is closely connected by ties of intercourse and common interest. It is the most Malayan of all these States; but its Malay Râja is, like the rest, responsible to the *Chow Kun*, or Governor of Sênggôra, and must look to be confirmed by the King of Siam. Kôta Bhâru, some miles on the east side of the Patâni river, is his residence; and the population of the country is to be found chiefly in this neighbourhood and near the upper valley of the Pêrak, which river the Rêman people use for exporting tin, &c. The boundary with Pêrak, near which are the valuable tin mines of Kroh and Intan, already mentioned in connection with Kêdah, has yet to be determined.

Jala, situated principally to the north-west of the river Patâni, near the head-waters of the Pêrak, lies under the eastward cliffs of the bold range of Bukit Bêsar. Jala is believed to be one of the richest mineral countries in the whole Peninsula, having abundant galena, tin, and gold already worked at some points by the Chinese. Like the other mineral countries, it is intersected by remarkable limestone formations.

The galena mines of Patâni, which a few years ago attracted much attention in Singapore, lie near the small town of *Banista*. This is situated in a picturesque amphitheatre of hills, through which the river flows, about 45 miles distant from the town of Patâni in a straight line, but double that distance by river.

In 1786, the year of the first Siamese invasion, there were said to be 115,000 inhabitants in the State of Patâni. In 1832, after the second invasion, there were computed to be only 54,000 people in these provinces, and the population has probably not increased since then, except in regard to its Chinese miners, who now number several thousands.

The southernmost of all the nine provinces collectively termed "Patâni" is Sai, beyond which lies the large and important Malayan State of Kêlantan.

Kêlantan.—South of Patâni on the east coast lie the Malayan States of Kêlantan and Trenggânu, whose position is one of independence guaranteed by treaty with the British Government, though nominally under some sort of subordinate relation to Siam.

Kêlantan is situated to the south of the Patâni States, the river Banâra being the boundary, between 6° 20′ and 5° 40′, with 60 miles of coast on the east side, and an area of about 7000 square miles; but so little is known of the interior that there is no great certainty about either area or population. It is in a prosperous condition, surpassing in population all the Native States on the east coast, and in natural resources and mineral wealth vying with Pahang.

It is bounded to the south by Trenggânu, the river Bĕsut separating them. It has the States of Rěman, Pěrak, and Pahang to the west, the eastern ridge of the peninsular chain being considered the boundary. The interior is believed to have a great extent of open country, traversed by the long but shallow river Kělantan **and its** tributaries, which, like the river Patâni, flow north. **Here there** is grown an immense quantity of rice, some of which **is** exported to Singapore ; cattle also are kept in large herds.

The town of Kělantan is situated near the river's mouth, and is a large and flourishing settlement with considerable trade. Its population is said to be over 20,000 ; and that of the whole state is estimated by the natives at 600,000, and on good authority as high as 200,000.

Its mineral resources comprise tin and gold. Even so far back as **1837** it was stated that 3000 pikuls of tin were exported annually, and **that** Kělantan gold, next to that of Pahang, was the most celebrated among Malays. Lead is also supposed to exist. Much pepper and other articles of export are also cultivated here by the Chinese, and a good deal of jungle produce is collected. The principal trade is with Singapore, and is mainly conducted by the Chinese during the south-west monsoon.

Kělantan is known to have existed as an integral State at the close of the 15th century and before the arrival of the Portuguese ; and in the Malay Annals it is specially stated that during the time of Mahmud II., of Malacca, A.D. 1477, Kělantan was a kingdom "more powerful than that of Patâni." Like Trenggânu, Kědah, and Patâni, it has, from time immemorial, been harassed by the demands of Siam ; and, according to the official statement of Mr. Anderson, Political Agent in 1825, **it** repeatedly solicited, in the early days of Pinang, the protection of the British Government and the establishment of an English factory, offering very considerable advantages. It has never submitted to Siam further than that, although practically under its own Malay Râja, it has made a customary acknowledgment of inferiority by periodically sending to Bangkok a tributary token called "the gold flower."

In 1832, the chief of Patâni, upon the invasion of his country by Siam, fled to Kělantan, but was delivered up to the Siamese *Prahbang*, who repeatedly summoned the Râja of Kělantan into his presence. With these mandates the Malay chief did not deem it prudent to comply, but was eventually compelled, it is said, to propitiate his foe, by a large present of specie and gold dust. Newbold pointed out at the time that this was a violation of the 12th Article of Major Burney's treaty of 1826, which stipulates that "Siam shall not go and obstruct or interrupt commerce in the States of Trenggânu and Kělantan. English merchants and subjects shall have trade and intercourse in future, with the same facility and freedom as they have heretofore had ; and the English shall not go and molest, attack, and disturb those States upon any pretence whatever." What little trade and intercourse now exist have passed from the hands of English merchants to those **of** Chinese and native traders.

Trenggânu is situated **at the** widest **part** of the Peninsula, between 5° 40' **and 4°** 35' N., and has an area of under 4000 **miles, with a population of** 20,000. Trenggânu has, for some time

past, included Kĕmãman, which lies to the south along the coast of
the Gulf of Siam. Its coast-line extends along the Gulf of Siam
for 80 miles, and is bounded on the south and west by the
principality of Pahang, and on the north and north-east by that
of Kĕlantan. The river Bĕsat is its boundary with Kĕlantan, and
the river Chĕrãting with Pahang. To the interior, the high ranges
forming the east boundary of Pahang form a natural frontier, but
the boundary is believed to be otherwise quite undefined. Of its
area nothing certain is known; nearly the whole country is one
continuous jungle, with less development, either of its minerals or
its commerce, than perhaps any other of the Malay States. The
inhabitants consist almost entirely of Malays and some wild tribes,
with a very few Chinese, who carry on the little that is now done in
the way of trade or mining. The total population of the State was
computed at 37,500 in 1856. Of this number, the town of Trĕnggãnu,
situated in the northern part of the State, near the mouth of a not
very large river, latitude 5° 25′ N., longitude 103° E., was
then estimated to contain from 15,000 to 20,000 inhabitants, or
more than half of the population of the State. A most destructive
fire took place in August 1883, which is said to have destroyed
nearly 2000 habitations. The place is but little visited, and the
small quantity of gold and tin produced come, it is said, from the
Pahang mountains, which are not so far removed from the coast at
this point. This State is claimed as one of the hereditary tributaries
of Siam, but has always resisted to the utmost the assumption of
any authority by the Siamese, the population being almost entirely
Mahomedan and Malayan. A "gold flower" is sent periodically to
Bangkok, through Sĕnggóra, in token of its nominal dependence,
but it has a treaty right to independent and unrestricted trade with
the British.

Kĕmãman (river and district) was, according to Malay tradition,
formerly a province of Pahang, and, on this ground, still considers itself
free from even nominal allegiance to Siam. This recognition is, however,
admitted by Trĕnggãnu, with which country Kĕmãman seems to be now
politically incorporated. It is a place of no importance, lying midway
between Pahang and Trĕnggãnu. The town is only a mile or two from
the mouth of the river of the same name, in lat. 4° 15′ N. It is a
settlement of modern origin, and probably owes its existence to the tin
mines, discovered early in the century, in the neighbourhood. The district
is scarcely 1000 square miles in area; and is, or until recently was, under
the direct control of a separate chief, under Trĕnggãnu. Its population
was estimated in 1839 at 1000 Malays and Chinese. It produces tin, a
little gold, camphor, ebony, &c. According to a Mr. Medhurst, who
visited the place in 1828, Kĕmãman at first yielded a considerable revenue
to the Sultan of Trĕnggãnu, but afterwards the mines failed, and the

Chinese dispersed. It is believed to be scarcely more prosperous at the present time than it was in 1859.

Between the rivers Kĕmáman and Trenggánu lie the smaller districts of *Paka*, *Duñgun*, and *Marang*, which, like Kĕmáman, are each under a chief, subordinate to Trenggánu.

SOUTHERN OR BRITISH DIVISION.

The British, as distinguished from the Siamese section of the Peninsula, is a purely political division, corresponding with no physical, ethnical, or other natural boundary. The two divisions are separated merely by an arbitrary or conventional line drawn from Kĕdah Peak on the west coast to the north frontier of Pahang, and running with the southern boundaries, whatever they may be, of Kĕdah, Rĕman and Trenggánu. But the northern limits of the Malay race lie considerably to the north of this line, south of which the administrative interference of Siam is scarcely at all permitted.

The British division, which, excluding the Straits Settlements, is even more thinly peopled than the Siamese, comprises five distinct political groups, as under :—

1. The three protected States of Pĕrak, Sĕlángor, and Sungei Ujong, occupying the west coast from Kĕdah to Malacca.

2. The so-called "Negri Sĕmbílan" group of petty inland States behind Malacca.

3. Pahang, on the east coast.

4. Johor, comprising the whole southern extremity.

5. The British colony of the Straits Settlements.

1. THE THREE PROTECTED STATES.

This group completes the British administration of the west coast from Pinang to Malacca. It comprises the Malay States of Pĕrak, Sĕlángor, and Sungei Ujong, ranking in importance in the order in which they here stand.

Pĕrak.—This State stretches for nearly 100 miles north and south between 5° 10'—3° 15' N. lat., and 100° 22'—102° E. long., and for probably a somewhat less distance inland, with an area approximately estimated by Mr. Deane at 7900 square miles, excluding the disputed district beyond Bukit Naksa. It is bounded north by Province Wellesley (Trans-Krian district) from Sungei Bákau on the coast to Párit Buntar, the boundary thence running to the source of the Krian river, which forms the frontier line towards Kĕdah. Between the Tăsek swamp or lake (the northernmost point at which

Pêrak and Kêdah touch), and the source of the Krian, the mountain range forming the water-parting of the Pêrak river constitutes the western frontier. From Tûsek eastwards Pêrak is conterminous on the north and east with the Patâni States of Jalo and Rêman; but here the exact course of the boundary-line is at present the subject of negotiations with Siam. On the east the inner range forming the divide of the Plus, Kinta and Kampar, Bâtang Padang and Bidor, with the other tributaries of the Pêrak, constitutes the frontier of Pêrak as far as the sources of the river Slim. On the south it is bounded **by** the river Bêrnam, and thence to Sungei Bâkau by the sea.

For commercial purposes Pêrak enjoys the advantage of proximity to Pinang, which is at present its chief outport, and with which **daily steam** communication is maintained. The chief harbours are the new Port Weld, Telok Kertâng at the mouth of the Lârut river, and Telok Anson on the Pêrak river.

The surface is almost equally divided between hill and plain, an extent of about 2000 square miles being occupied with uplands ranging from 1500 to 8000 feet above sea-level. The chief mountains are the Titi Wangsa and Gûnong Hîjau (Lârut), the Gûnong Bûbo, and the inland ranges, of which Gûnong Besar, Mounts Robinson and Tengah (Tangga) are respectively the central peaks. Here rise the **rivers Plus,** Kinta, Kampar, Bidor, **Songkei, Slim, and the** Sembîlan, **flowing** to the Pahang.

The mountain ranges are generally of granitic formation, but, in strong contrast to their usually round surfaces, sharp peaks and crags of limestone formation crop up here and there throughout the country. The principal of these are Gûnong Kênderong, Gûnong Kernei, and Bukit Kajang in **the** north; Bâtu Kurau and Gûnong Pondok; some unnamed hills in **the** Plus ranges, and numerous peaks in the Kinta valley.

The caves in the limestone mountains furnish bats' guano—an excellent **manure,** which, as well as lime, is available for both mountain and low **country** cultivation.

The *Residency.*—The seat of the Government of the British Resident is **the** small village of *Kuâla Kangsar,* on the upper waters of the Pêrak, about 23 miles from the port of Teluk Kertang, on the Lârut river, with which it **is** connected by a good road. It lies about 100 miles up the Pêrak river, the Lârut route thus giving the most direct access to Pinang.

The country can best be described as consisting physically of three principal water-systems—that of the Krian to the north, that of the Pêrak in the centre, and that of the Bêrnam to the south. Each will be described in turn. But the tin-mining district of Lârut, which belongs to neither of these physical divisions, has played and still plays so important a part in the development of the State, that it deserves first mention.

Lârut is situated about midway between the river Krian and the river Pêrak.

For about thirty years, Chinese miners have worked the extensive tin deposits of great richness at the base of the high mountain range called Gûnong Hîjau, and on each side of a small river called Sungei Lârut. This place was found by the early pioneers to be not only rich in tin, but most advantageously situated in respect of commercial intercourse with the British port of Pinang, some sixty miles distant. Tin-deposits are rarely found so near the sea as in Lârut, which is under the immediate charge of an Assistant Resident.

Thaipeng, the principal town, and the largest place on the west coast, Malacca not excepted, is the centre of the mining industry, and is about eight miles from the sea-coast. It is the head-quarters of the chief departments of the State, and is connected with Kwâla Kangsar by a carriage-road, and by a line of telegraph. The main road to the sea has hitherto been from Thaipeng to Têluk Kêrtang, but a short line of railway, intended to connect Thaipeng with Port Weld (eight miles), is now completed, as well as a road from Lârut to the Krian river, which will open up communication by land with Province Wellesley. There is also telegraphic communication with Pinang.

Owing to the proximity of the mountains to the sea in this part of the Peninsula, the rainfall in Lârut is heavier than elsewhere along the Straits, amounting in 1884 to 166 inches.

The Krian Basin.—Krian is an agricultural district adjoining Province Wellesley, the seat of extensive sugar and rice cultivation. It has a large Malay population, consisting principally of settlers from Pinang, Province Wellesley, and Kêdah. A good many Chinese and Tamil planters have also recently settled there.

Sêlâma, 70 miles up the Krian river, on a large tributary of that name, forms a tin-mining settlement, which a few years ago was more flourishing than at present. It is situated near the principal bifurcation of the Krian. There is a colony of Sumatran Malays at Sêlâma, besides Chinese miners.

The Pêrak Basin.—The Pêrak, perhaps the largest, and certainly the most important, river on the west slope, drains not only the extensive valley of the State to which it gives its name, but also receives the drainage of the considerable Kinta district, comprising together at least half the area of the State. It is navigable for small steamers as far as Têluk Anson, the capital of Lower Pêrak. Its source is said to be in the frontier mountain Jambul Mêrak, from which the Têlûpin and Patáni also take their rise. Its whole length is about 250 miles. At first it flows in a southwesterly direction towards the sea, receiving, from the west, the

Rui, the Kěnderong, and the Kěnĕring; and from the east, the Sengo and the Těmŭngan. From Kwăla Kěnĕring to Bandar its course is due south, and all its main affluents flow into it from the east. Such are the Plus, the Kinta at Kôta Lûmut, the Bâtang Padang, and Bidor at Dûrian Sabâtang, and near its estuary the small river Sungei Jandarâta, which almost connects the streams of the Pêrak and Běrnam rivers, here flowing parallel at no great distance from each other. The Pêrak empties itself into the Straits, a few miles to the south of the Dindings. It has a wide estuary, but here, as in other rivers in the Peninsula, shallow water on the bar at the mouth impedes navigation. The principal places on this river are:—Kôta Setia, Têluk Anson, Dûrian Sabâtang, Bandar, Kôta Lûmut, Bandar Bhâru (the former Residency near the junction of the Kinta), Pûlau Tiga, Lambok, Bôta, Blanja *Kuâla Kangsar* (the present Residency), Sâyong (the residence of H. H. the Regent), Kôta Lâma, Chegar Gâlak, and Kôta Tampan.

Tin is found almost throughout the valley, but in greatest quantity near the east bank of the Pêrak and in the Kinta district. The Kinta district includes the territory watered by the river of that name and its tributaries. A Collector and Magistrate has charge of it, and resides at Bâtu Gâjah on the Kinta river. Other places of importance in the district are Lahat, Pâpan, Ipoh, Pěngkâlan Pĕgŏ, Kôta Bhâru, Pěngkâlan Bhâru (Sungei Ray), Gôpeng (a large Chinese mining settlement), Kampar, and Chěnderiang.

The Běrnam Basin.—The southernmost **district of the state is** that of the river Běrnam, probably the **largest river, in** regard to volume of water, to be found in the **Peninsula.** It is about **two** miles wide at the mouth, and navigable **for large** steamers for many miles. Though draining a very different district, **its** mouth is less **than** twenty miles from that of the Pêrak.

Proceeding up the Běrnam, almost due east, the chief places (though **none** of them are of any size) are *Sâbak*, about 20 miles from the mouth; *Telok Kurâli*, about 73 miles from the sea, where the river is still about 120 yards wide and very deep; *Changkat Bĕrtam*, 85 miles by river from the sea, a small rising ground planted with durian trees, and occupied by a colony of a few Malays. Above this spot stretches an immense expanse of unhealthy swampy country for miles on both sides of the river. Through this swamp the Běrnam winds down from *Gĕdeangsa*, 111 miles by river from the sea, where the land again becomes higher. A series of canals or cuttings, shortening the navigation of the river, and making it available for steam-launches, have recently been made from this point, through the Changkat Bĕrtam swamp. The distance for boats is, it is computed, thus reduced from 111 to about 50 miles.

Kuâla Slim, about 130 miles up the river, was formerly the principal station and the Collectorate of the district. It is situated at the bifurcation of the main stream, where it divides into two branches of similar size

—the Slim running down from the direction of Pérak in the north-east, and the Bérnam from Sëlängor in the south-east. A hilly region called Changkat Lela divides these branches.

Ulu Slim lies about 30 miles higher, at the confluence of the Slim and Gëliting. It is described as very picturesque—"it might almost be in Switzerland." From here there is a short overland path to the Pérak waters (river Songkei); and some of the nearest affluents of the Pahang river, flowing into the Gulf of Siam, have their source in the same mountains, which are the source of the Slim branch. The watershed of the river Bérnam, which flows from the south, is to be found near the Sëlängor-Pahang boundary. At Ulu Slim land has been successfully opened up by English coffee-planters within the last few years.

The highest station on the Bérnam river is Ulu Bérnam (Tanjong Mälim), a fertile, well-cultivated station at the foot of the dividing range. The main road to Sëlängor and Bérnam passes through Tanjong Mälim, which has quite recently been made the head-quarters of the Bérnam Collectorate. A trunk-road from Kwála Kangsa to this point will soon be opened. Here it will join the road recently made by the Sëlängor Government, thus completing an unbroken highway of nearly 300 miles from Malacca to Butterworth in Province Wellesley.

Trade of Pérak.—The chief export is tin, amounting, in 1893, to £400,000; and the abundance of this metal is the most important economic feature of the State at present. The other exports amount to £700,000 (including sugar, £61,000); and the whole trade, imports and exports, now exceed £2,100,000. There is now daily communication by trading steamers between Pinang and Lárut. steamer also touches at Dúrian Sabâtang on her fortnightly voyages between Singapore and Pinang, and there is a separate service between Pinang and Téluk Anson. There is also regular steam communication between Pinang and Bérnam.

Government.—The government is carried on under the Rája Mûda, as Regent, aided and advised by a British Resident, and a Council consisting of the Resident and Assistant Resident, and Native Chiefs of rank and influence.

A military police force of 700 men, mainly Sikhs, is maintained to secure order, with half a battery of Artillery.

The Collectorates are at *Lárut*; at *Pârit Buntar* and *Sëláma* (for Krian); *Kwála Kangsar*, the seat of the Residency; *Téluk Anson* (for Pérak river); Bâtu Gâjah (for Kinta district); Ulu Bérnam (for the Bérnam).

History.—Pérak is one of the oldest States in the Peninsula, and its history has been maintained with scarcely a break for 300 years. It was subject to Achin in the days of the Portuguese, and until the close of the 17th century, but otherwise it appears to have maintained its independence throughout. It was overrun and occupied by Kédah during the Siamese hostilities in 1821; but the invaders were induced to withdraw by agreement with the Pinang Government in 1825. The Dutch had continually tried, with varying success, to maintain a trading monopoly of the tin for 150 years, but their attempts to obtain a footing were unsuccessful, and all European interference with Pérak ceased until 1818. In consequence of the cession of Malacca to the Dutch in that year, the Pinang Government entered into commercial treaties with Pérak, among other Native States, in order to forestall any fresh attempts at Dutch monopoly.

This alliance proved useful to Pérak a few years later, when the Siamese attempted to overrun the country, but were checked by the British authorities at Pinang.

The recent development of events dates from the rise of Lárut into importance under Ngah Jafar, in 1852, consequent upon the discovery of the rich tin deposits there. The Chinese then came in great numbers, and before long the Malay Government naturally fell to pieces. After some years of anarchy, Governor Sir Andrew Clarke interfered in January 1874, and the Pangkor Treaty was made, introducing the "Protected States" experiment. The small rising that brought upon Pérak a military occupation, after the assassination of the first Resident, Mr. Birch (1875), led to the adoption of the more robust policy under which Pérak has made its subsequent rapid advances. The State has quite recently opened the first railway in the Peninsula.

Sëlángor.—The Protected State of Sëlángor adjoins Pérak along its whole southern frontier. It is situated between the parallels of 3° 45' and 2° 40' N., with a rather greater extent of coast-line on the east shore of the Straits than its northern neighbour Pérak. Area between 4000 and 5000 square miles. Population about 50,000.

Sëlángor is separated from Pérak by the Bërnam river, which forms its northern boundary. Its extent along the coast is about 100 miles, as far as the river Nipah to the south (since the 1877 boundary was fixed), and then by a line drawn in 1884 to the north and east as far as the hills which divide it from Jëlëbu.

Sëlángor is a comparatively recent State, the western part of its territory having apparently been left unoccupied from time out of mind, to a greater degree than any other portion of the Peninsula. The southern division was formerly a separate State—Klang (Këlang) —one of the four original States of the "Nëgri Sëmbïlan" confederation.

Under the name of Sëlángor are included four main districts, each having a considerable river, named respectively Bërnam, Sëlángor, Klang, and Langat; Bërnam to the north, and the others further south in the order in which they are named. With the exception of Klang and the mouth of the river Sëlángor, the whole territory of the State was absolutely *terra incognita* until quite lately. Lúkut, now comprised in the Sungei Újŏng frontiers, was formerly part of Sëlángor. Being rich in tin found close to the shore, and being situated at a distance of only 40 miles from Malacca, this district was, under a former Rája, the most thriving in Sëlángor.

The greater part of Sélângor is an extremely level country, stretching inland about 30 miles in the south, and nearly 50 miles in the north, and as yet but little cleared and very thinly inhabited. In its wide versant it presents a marked contrast to Pérak, its rivers flowing almost due west instead of southward. In the interior are some high spurs thrown out from the great mountain chain, especially between Ulu Sélângor and Ulu Langat, and in the neighbourhood of Kwâla Lumpor, the present capital. These spurs have an altitude of about 2000 feet, with numerous high peaks, where they join the chain, reaching to more than 5000 feet. The highest is Bukit Téngah (6200 feet), in the Gûnong Kâli spur.

At Genting Bidei, 22 miles north-east of Kwâla Lumpor, there is a pass into Pahang at the junction of two important spurs, one running due south behind Kwâla Lumpor, the source of the river Klang; the other trending away inland, leaving a valley which widens to about 10 or 12 miles, down which flows the Ulu Langat. Several of the highest peaks in this group rise to above 5000 feet.

Further north, the river Sélângor rises among even higher summits in the central chain, which is here at its nearest point to the river Pahang, whose tributaries flow down to the east from the same hills. The highest peaks after Bukit Téngah are:—Gûnong Râja, 5450 feet; Gûnong Chim-bêrax, 5650 feet; Gûnong Péchêras, 5650 feet; and Bukit Kanching, from which rises Sungei Bâoh, south of Sélângor.

In Sélângor the chief towns are:—

Kwâla Lumpor is, and has been for many years, the centre of the tin-mining of the country. In 1879 it was made the capital, instead of Klang. Its distance from the nearest navigable waters (24 miles) is its principal drawback; but it is well placed for inland communications. The track distinctively known as the "Pahang Road" runs due east from Kwâla Lumpor, which will soon be connected by rail with Klang. Klang, the principal port of the country, 12 miles up the river, former seat of Government and the first Residency. It is situated near the sea, and many miles from the vicinity of the tin-mines at the foot of the mountains, but is favoured with a navigable river which, owing perhaps to Kalang island lying across its mouth, is without the almost invariable bar. Sélângor, lying at the mouth of the river Sélângor. The river is shallow and practicable only for vessels of small burden. The Dutch had formerly an establishment at Sélângor for the monopoly of the tin; and a stone fort of their construction is still a conspicuous object, having formerly been, next to Malacca, the most formidable stronghold in these waters. The other towns of Sélângor are Langat, Bandar Kanching, Jugra (where the Sultan resides), and Ulu Langat. This latter lies more inland than any other part of the State.

A good bridle road is now completed from Berânang to Ulu Bérnam, connecting Sungei Ujong with Pérak by means of a main road through the whole length of Sélângor from south to north.

Population.—The native inhabitants are believed to be the descendants of a colony of Bugis, from Goa, in Celebes, who settled here and at Kwâla Linggi under their Chief, Aron Passwai, towards the commencement of the last century. The population about ten years ago had fallen away to a minimum, in consequence of the incessant quarrels and misrule of its princes. It has been much increased of late years, both by Chinese settlers and miners, and by the immigration of Malays from less prosperous States in its neighbourhood, including not a few from Jambi and other places in the east of Sumatra.

Products.—Sêlângor produces tin of excellent quality, and the deposits at Ulu Langat and Kwâla Lumpor have proved extremely rich, the latter under the name of Klang tin having attracted much attention for the last twenty years. For some years past, it has stood second only to that of Lârut. The mining is carried on almost entirely by Chinese. The mines opened by European enterprise in 1883-4 have all been unsuccessful except that at Râwang, in Ulu Sêlângor. Besides tin, there is little else but jungle produce, though important plantations of coffee, pepper, sugar, &c., have been commenced.

Government.—An incessant quarrel, chiefly as to the rights over the tin duties levied in Klang and Sêlângor, prevailed from 1867 to 1873. At the time when Governor Sir Andrew Clarke was settling the affairs of the Native States in 1874, he undertook to assist the Government of Sêlângor. The Government of this State has since been carried on under the same system as Pêrak.

Mention has been made of a railway to run from Kwâla Lumpor to Klang, and roads will shortly connect Sêlângor with Pêrak, as it is already connected with Sungei Ujong and Malacca. It may confidently be expected that a country with such mineral resources, and such fine hills and plains, drained by abundant rivers like the Bérnam, Klang, Sêlângor, and Langat, will, under a peaceful rule, become populous and wealthy.

Sungei Ujong, the smallest of the three Protected States, lies to the south of Sêlângor, between that State, Jêlêbu and Rêmbau, to the north-west of Malacca. It was one of the four original States out of which grew the "Nêgri Sêmbîlan." Including the districts of Lûkut and Sungei Râya, it has an area of about 500 square miles, mainly on the north bank of the river Linggi. Sungei Ujong suffered for many years from the Sêlângor disturbances, which owed their origin to the same cause—to quarrels over the tin-royalties. But Sungei Ujong has always been, especially since the development of its mines, the leading State of the Nêgri Sêmbîlan.

The Linggi river, its one large stream (the highway to Sungei Ujong and much of Rêmbau), had, in 1873, been rendered impassable by constant border fights between these two States. After repeated complaints on the part of the British subjects in Malacca of the violence and extortion that put a stop to all traffic on the river Linggi, Sir Andrew Clarke, Governor in 1874, went personally to Sempang on the Linggi river, and re-opened trade and suppressed disturbance. A Residency was established in Sungei Ujong shortly after, to prevent further disturbance, and to protect the large number of Chinese miners working there.

The mountains of Sungei Ujong approach the sea more nearly than those of Sêlângor, the interval being, however, even more uncleared and swampy than in the northern State. In former times, Sungei Ujong seems to have been a wholly inland State, but since Residents have been stationed in Sêlângor and Sungei Ujong, the frontier line between them has been modified; and now the river Lûkut and district, formerly renowned for its tin, but since 1860 almost deserted, are included in Sungei Ujong, thus giving it 20 miles of coast, between Sungei Nîpah and Kwâla Linggi.

The tin workings, and the best inhabited portion of this small State, lie in a sort of semicircular valley, between the hills Berembun (4000 feet), in which the Linggi rises, Taugga (1800 feet), the Jélébu boundary, and Perhentian Rimpun (2000 feet), at the Sélángor boundary. Mount Berembun is, in some respects, the key to this State, and, it may be said, to the whole of this part of the country. On the south side of this mountain flows the Moar, and on the east a feeder of the Pahang river—Triang.

Through a gap called Bukit Pûtus, between this mountain and Gûnong Angsi, to the south, is a pass leading to Sri Mènanti and the other Negri Sěmbílan States.

Population, Products, &c.—The Malay population is almost entirely agricultural, and is mostly found near the mountains, as at Pantei. The whole population is probably below 14,000 souls. Sungei Ûjong has abundance of water, and its land is considered suitable for the cultivation of coffee, cocoa, cinchona, &c., which are being grown both on the hills and plains. On the lowest ground, tapioca is now cultivated. Tin-mining is still carried on to a considerable extent by the Chinese at Ampangan, near the Residency, and its neighbourhood. These Chinese miners in Sungei Ûjong, as in Lârut, have been the real sinews and wealth-producing power of the country.

A road now connects Sěremban with Pěngkálan Kempas, the newly-opened port near the mouth of the Linggi, and there is regular communication by steam-launch between Malacca and the Linggi. Not far above Pěngkálan Kempas are Permátang Pásir, the former inconvenient " port," and Linggi village. The Residency is at Sěremban, about 22 miles higher up the country. Two miles nearer the port is Kásu, the Customs' station at the bridge over the Linggi river ; to which place the stream is, or was before this new road was made, navigable and clear for small boats. From Sěremban a road has been made through Pantei to Sélángor on the one side, and Malacca on the other ; and upon this the first instalment of the future road up the Peninsula, Sěremban, and Sungei Ûjong generally, have a fine central position.

Government.—The residential system was introduced here shortly after its adoption in Pérak and Sélángor (December 1874), and with a short break, at the time of the Pérak war, that form of government has since been peacefully carried on in the manner already described.

2.—THE NĒGRI SĚMBÍLAN STATES.

These small States, formerly a kind of confederacy of Nine States, of which the name alone now survives, occupy about 2000 square miles of the interior between the Protected States on the north, Malacca on the west, Johor on the south, and Pahang on the east.

This confederacy has since Sir F. Weld's treaty of 1883 been under British protection, and roads have been made connecting Johol and Rěmbau respectively with Malacca.

Apart from Klang, which has long formed part of Sélángor, and Sungei Ûjong, which, as a separate Protected State, is now on a different footing, the Négri Sěmbílan contain a total population of not more than 30,000, mostly to be found in Rěmbau and Sri Měnanti.

Originally there appear to have been four States, which were afterwards broken up and modified as shown below :—

Formerly.	At present.
Kelang.	—
Sungei Ujong.	—
Jělěbu.	Jělěbu.
	Johol.
Johol	Inas or Jělei.
	Ulu Moar or Sri Měnanti.
	Jěmpol.
	Rěmbau.
(Four States.)	(Six States.)

Of these six States, those of sufficient importance to deserve special description are Jělěbu, Johol, Sri Měnanti and Rěmbau. They had their origin most probably in the organisation introduced by the Měnangkábau tribes, who have emigrated at different times from Sumatra to this part of the Peninsula. In the days when Johor was powerful, the Něgri Sěmbilan were under the Sultan of Johor ; but about 1773, Johor being indifferent about the government of these remote tribal States, allowed the Dutch to obtain for them, at their request, a Prince of true Měnangkábau descent (Rája Melewar), who, as Yang-di-Pertûan Běsar, ruled over the Confederacy. The States were thus formally federated, each retaining, however, its own Pěnghúlu or Dato'. The real power is vested in the Pěnghúlu, that of the suzerain being nominal only.

This Sumatran immigration, and the political intercourse of the independent Princes of Sumatra with those of the Peninsula, deservedly attracted the attention of scholars like Marsden, Leyden, and Raffles : but the whole arrangement was of too artificial a kind to last long, and after five successions of Měnangkábau Princes, they ceased to be invited over (1820). It is noteworthy, however, that even the more civilised Malays, especially in Rěmbau, still hold to the tribal organisation : the very names of many of their tribes, such as " Anak Acheh " (children of Achin) and " Sri Lěmak Měnangkábau," betraying their comparatively recent migration from Sumatra.

Jělěbu is a small State lying to the north and east of Sungei Ujong, and containing about 400 square miles, and under 1000 inhabitants. It belongs *politically* to the west **coast**, though *physically* to the east coast. It has thus a peculiarly central position in regard to this region of the Peninsula, being situated at the great water-parting of the southern portion of it. Jělěbu had, until the year 1884, remained unexplored. It lies between

Sungei Újong and the valley of the river Pahang, having Sélángor
to the north and Jémpol to the south. The country is a succession
of narrow valleys between hills of no great height, except in the
south, where they culminate towards Gúnong Berembun. These
hills are the sources of many of the principal rivers on both sides
of the Peninsula—the Linggi and the Moar flowing to the west,
and the Serting and Triang to the east, both feeders of the Pahang.
Génting Pireh is the boundary towards Sélángor. It is about 28
miles from Úlu Langat, and not far from the mining settlement at
Sungei Lui. Bukit Tanggra (1800 feet), at the head of the Kláwang
valley, lies between Jélébu and Sung i Újong, and deserves notice
as the furthest western point of the east coast watershed. Jáwi-
Jáwi Bétáub, on the Triang, is claimed by Jélébu as the eastern
boundary towards Pahang, but this has still to be settled. Meanwhile Sungei Túa has been adopted (1884) as the provisional
boundary. At the point where Sélángor, Sungei Újong, and Jélébu
meet is the hill *Perhentian Rimpun*, said to be so named from
the assembly (Berhimpun) of the Chiefs of the old "Four
States."

Hitherto communication has been maintained chiefly with Sungei Újong,
a brídle-path connecting Séremban with Jélondong, centre of the mining
districts. Some parts of Jélébu will probably be found most accessible
from Sungei Lui in Sélángor, while others may be more easily approached
from Malacca by way of the valley of the Langkap, one of the head-
waters of the Úlu Moar, which runs down the southern side of the
Berembun towards Téráchi. The geology and physical geography of this
state alone are of any present consequence.

The only industry, beyond the cultivation of a little rice chiefly in the
Kláwang valley, is some tin-mining carried on by Chinese at Jélondong,
near the Triang and close to the Pénghúlu's place, Kwála Glámi. The
tin-deposits lie on the Pahang side, and are said to be easily worked.

The Triang, of which the head-waters may almost be said to form the
State of Jélébu, is an important feeder of the Pahang, and both the main
stream and its largest tributary (the Kénábui) are deep and navigable for
most of the year. Rice is thus imported easily from Pahang. The tin-
deposits in Kénábui, Jélondong, and Kwála Glámi, are unusually rich.

The State has always been one of the Négri Sémbilan, ruled like the
rest by an elective Dato' Pénghúlu, with a Yam Túan whose only function
seems to be to represent the hereditary and monarchical principle. A
Collector and a small detachment of Sungei Újong police have recently
been established at Jélondong.

Johol, which formerly included the whole country to the interior
of Rémbau, Malacca, and Ségámat, is now broken up into the
separate States of Jélei or Inas, Sri Ménanti, and Jémpol.

The four Bátins, or aboriginal chiefs, were those of Klang,
Jélébu, Sungei Újong, and Johol (see p. 31).

According to the natives, the former boundaries of Johol were Mount Ophir (*Gûnong Lédang*), and from there Kantau Pait on the Mûar a little above Kuâla Pâlong (towards Johor), thence to Lûbok Serampang on the Serting (towards Pahang), thence to the Jélébu boundary at Jâwi-Jâwi Bétâub on the Trîang, and thence to Sungei Langkap in Ûlu Mûar, and along Gûnong Brimbûn (towards Sungei Újong) to Bâtu Gâjah in the Pâbei pass (towards Rêmbau).

The present State of Johol, which has little political importance, is an undulating country without either large streams or high hills, and though known to contain much gold, especially on the Gémas (*Sungei Mas* or gold river?), there are no workings at present. Its chief or Pénghûlu resides at Kwâla Tûmang.

One of the principal districts is Inas or Jélei, at one time perhaps a separate State of the Négri Sémbilan, with which it may now be considered to be incorporated.

The Johol and the Inas both flow into the Jélei, which falls into the Mûar. The lower part of the Jélei is claimed by Johol, so that it is a sort of little Switzerland, enclosed by Rêmbau, Sri Ménanti, Johol, and Tampin. The direction is south-east of Sri Ménanti.

Johol has been ruled by its Pénghûlu, Dato' Eta, for over forty years, but since the arrangement of 1876 he has been under the nominal authority of the Yam Tûan of Sri Ménanti.

Sri Ménanti, as recognised by the British Government in 1876, contains about 300 square miles, and a population of about 3000. It is the old State of Ûlu Mûar with the addition of Jémpol to the east. The open valleys of Bandul and Térâchi, watered by the upper Moar, lead from Bukit Pûtus, the frontier of Sungei Újong, to Sri Ménanti. This was formerly the seat of the Yam Tûan or Ménangkâbau Prince, whose titular pretensions formed a bond between the Négri Sémbilan free States.

The country is chiefly flat, with some hilly districts **about the** sides of Gûnong Pasir and *Perhentian Tinggi*, which is the natural boundary towards Rêmbau. The pass across it, connecting the two States, is about 1150 feet high.

Sri Ménanti is tolerably prosperous, though, as in all the Négri Sémbilan, its rice-crops have, for many years, been deficient. About 500 Chinese carry on tin-mining at Bêting and Kuâla Pîlah. Sir F. Weld, who visited the district in 1885, found the Mûar river at Kuâla Pîlah, probably over 150 miles from its mouth, still 20 yards wide. But some obstructions having been placed in the stream, boats no longer ascend to this place. A few miles lower down the Mûar is separated only by a narrow space from the head-waters of the Sungei Hilir Serting, a head-stream of the Pahang river. Hence from its central position this district is of great political and commercial importance.

D

If Jělěbu is of more consequence in regard to its physical than its political relations, it is just the opposite with Sri Měnanti, the position of which is essentially political. It was without a Chief for some years before the treaty of 1876. But after our military occupation in 1876, and upon the withdrawal of our troops, the office of Yam Tūan, which seems to have been in abeyance, was re-established. By the treaty of that year, Tunku Antah obtained the administration of Sri Měnanti, with a general authority over some of the other small States.

Rembau, originally an offshoot of Sungei Ūjong, has an area of about 400 square miles. It is not only the best known, but is, in every respect, at present the most important of all these small States. Physically speaking, Rěmbau is but an extension of the plain of Malacca, with no natural boundaries, except at one or two points, to separate the two countries. In fact, until fifty years ago, the portion of Malacca nearest to Rěmbau, called *Nāning*, was itself an independent State.

The boundaries of Rěmbau are not very well defined. Those with Malacca territory are the places named in the Treaty of the 9th January, 1883, and the Rěmbau branch of the Linggi, above Sempang.

The boundary with Sungei Ūjong was fixed in 1881 as follows :— From Sempang to Bukit Mandi Angin, Pěrhentian Tinggi, and Gûnong Angsi.

The boundaries with Sri Měnanti are said to be Gûnong Tûjoh, and Gûnong Lîpat Kâjang, and those with Johol, Bâtu Gâjah and Gûnong Dato'.

The inhabitants are now mainly of Sumatran race, having immigrated principally in the 17th century. The almost exclusive industry of the country has always been padi-planting, for which its heavy rainfall is an advantage. In recent years, tapioca has been cultivated by the Chinese, which has materially increased the prosperity of its people. Tin is known to exist in some quantity, especially in the river Pědas, but the prejudices of the people have hitherto prevented mining.

The soil and physical configuration of Rěmbau generally resemble those of Nāning. The country is of an undulating character, the depressions being mostly planted with "sâwah," or wet padi-fields. Bukit Bôsar is the only mountain, exclusive of the frontier ranges towards Sungei Ūjong, Sri Měnanti and Johol.

Of the towns Sempang deserves first mention. Here the Rěmbau and Pěnar join and form the Linggi, and a Police Station stands in the angle thus formed, on some land ceded to Government in 1874. It was formerly one of the chief places in Rěmbau. Kuâla Pědas, a few miles up the Rěmbau on the right bank, was another; but both these districts have

been deserted. Nor is the capital easy to define, for each successive
Pénghúlu seems to have his own. Bandar Rásau was the residence of
the Yam Túan Múda, and latterly of the ex-Pénghúlu, Haji Sail. In
1837, Newbold said the Pénghúlu resided at Chémbong: the present
Pénghúlu resides at Gemayun near Chéngkau.

The Government of Rémbau is the best type of the tribal system to be
found in the Peninsula. In something like its present form, it probably
came over with the earliest immigrants from Sumatra, and has since been
maintained with great conservatism among the twelve Sukus or tribes. It
is by and among the *Lémbága*, or hereditary chiefs of these tribes, that
the Pénghúlu must be elected. This election follows very minute and
elaborate rules, grafted by the Sumatran immigrants upon the aboriginal
system, of one feature of which the following is a summary :—

"Béduanda is the name of one of the chief aboriginal tribes in the
south of the Peninsula, and two of the chief Rémbau clans bear the same
name—the Béduanda Jawa, and the Béduanda Jakun—from which the
Pénghúlu is alternately elected.

"This alternate election is said to be due to a dispute between the two
branches of the Béduanda, over the right to elect the Pénghúlu, which was
settled by the sovereign of Johor giving each the right alternately.

"At the same time, he gave distinctive titles to the Pénghúlus—to the
one elected from the 'Béduanda Jawa' that of 'Sédía Rája,' to him of the
'Béduanda Jakun' that of 'Léla Mahárája.'"

The office of Lémbága, or electoral chief, is hereditary, descending on
the side of the sister, as in Náning and all the Ménanagkábau States.

3.—PAHANG.

Pahang, between Trínggánu and Johor, extends along the eastern
side of the Peninsula from 2° 40' to 4° 35' N., and has about 130
miles of sea-coast on the Gulf of Siam. Its boundaries are the river
Chéráting, with Trínggánu; the river Endau, with Johor; and a
line along the eastern frontier of Jélébu, Sélángor and Pérak to the
west. To the north-west the boundary is not defined, but may
be taken as following the watershed of the Ulu Pahang.

Its area probably exceeds 10,000 square miles, and its line of
greatest length, from Ulu Endau to Ulu Pérak, approaches 200 miles,
far exceeding that of any other State in the Peninsula. Besides the
territory on the mainland, Pahang includes two chains of islets running parallel to its coast, generally at about 25 miles distance. The
State of Pahang, apart from these islands, is almost identical with
the basin of the river of the same name, in an even greater degree
than is the case with Pérak. This river is shallow and, therefore,
not the largest in volume; yet, as regards its position in the very
centre of the Malay Peninsula, and the extent of country it drains—
from 3° to 5° N.—the river Pahang may fairly be considered the
principal stream in the whole region.

Pahang is, in many respects, the least known, geographically and otherwise, of all the the Malay States, and it offers a most interesting field for exploration. Here are found both the highest mountains and the widest extent of lakes and marshes in the Peninsula.

The highest summit in the Peninsula is believed to be Gûnong Tahan, which has not been ascended, or even seen by Europeans except at a great distance, but which almost certainly reaches a height of between 10,000 and 12,000 feet. This central chain of the Peninsula at its widest point, is situated to the east of the upper waters of the Pahang, and can probably be best reached from the Ulu Temling, a feeder of the Pahang, near Jêlei. The geological formation of the hills consists, so far as is known, of granite, sandstone, shale, and clay. Some of the islands, as Tioman and Tinggi, consist partly or entirely of trap rock.

The next highest summit is to be found on the opposite side of the Pahang valley, in the neighbourhood of Gûnong Rãja, near the Sêlãngor boundary. Other high hills are found in the eastern chain, from which flows the river Chêrãting (called the Sêrting near its source), the Dûngûn in Tringgãnu, and the Lêbih in Kelantan; and the Gûnung Chem south of the Pahang, which is believed to supply the Chêno lakes. Still further south is the lofty Gûnong Gayong, source of the Rumpen.

The Chêno lakes, and the others in the neighbourhood, as, in fact, the water system generally, are peculiar to Pahang. The Pahang river drains a great length of country, and, in its course, receives important feeders from all directions—from the mountains to the north, south, and west. The lower part of the stream, below Kuãla Bêrã, flows for nearly 100 miles due east, through a very flat and marshy country. The river and its feeders here become wide and shallow, opening out into spaces like small lakes. The country between Pahang and Rumpen is particularly level, and the three main tributaries from that region—the Bêrã, the Chêno, and the Chêni—are all noted for such lakes. That of the Bêrã is the largest sheet of inland water in the Peninsula, but its shores, like the Chêni, are only inhabited by Sakei, while the Chêno lakes are inhabited by Malays.

The shallowness of the Pahang rivers makes them navigable for small craft only, except in the rainy season. Unlike those on the west side, their banks are sandy, often high, and mostly free from mangroves. The Pahang is formed by the Lipis and the Jêlei flowing respectively from the north-west and north, and uniting a few miles below the Penjum river. One day's journey lower down the main stream is joined by the copious Temelin from the north-east. Below the Temelin confluence it takes the name of Pahang, and before reaching the coast receives several other large feeders, such as the Semantan, Triang, and Bêrã, all from the west and south-west.

Like most of the eastern sea-board the coast of Pahang is mainly an uninhabited forest; but it has the advantage of a fine sandy shore with numerous Ru trees (Casuarina littorea), forming a natural highway, practicable even during the north-east monsoon. Such a coast route nowhere occurs on the west side, where the muddy foreshore is everywhere overgrown with dense mangrove jungle.

Pahang is far from being a populous country, even according to the low standard of the Peninsula, although there are a good many flourishing Malay settlements, especially in the interior. Here the valleys of the

Ranh and Lipis, tributaries of the Pahang from the Ulu Sĕlàngor hills, together with the Jĕlei and Temaling districts lower down the main stream, appear to be more thickly inhabited than any other part of the country.

Pahang owes its chief importance to its rich gold and tin deposits, which for productiveness and quality are unrivalled in the Peninsula. The chief gold mines are, or have been, in the Pahang basin at Lipis, Semantan, and Lŭet; and the same metal is also found as far south as the Bĕrà river. A mine of galena is found at Sungei Lembing on the Kwantan, and tin everywhere near the gold-diggings, and in places like the Triang and Bentong valleys where gold is not worked.

Recent explorations (1885) show the country to be very poor at the present time. The Lŭet valley is now almost depopulated, not more than fifty able-bodied men being found in the whole district. Here there is no auriferous quartz, although a great deal of surface-mining appears to have been formerly carried on, traces of extensive Chinese diggings having been found.

Of the "mineral" States the Malays rank Pahang first, Kelàntan next, and Patàni third; these territories alone yielding galena as well as gold and tin. It is noteworthy that the chief gold-workings lie almost entirely along a somewhat narrow zone running northwards from Mounts Ophir and Segâmat through the very heart of the Peninsula to the Kalian Mas (gold diggings) of Patàni and Telopin. The best tin-workings of Pahang lie on the river Bentong, and near the auriferous district of Jĕlei. In whiteness and pliancy the tin of Pahang rivals that of Pérak and Sĕlàngor on the west coast.

Pahang is said to grow sufficient rice for its own consumption, besides exporting a little to Jĕlĕbu. It is mostly wet rice, the buffalo being employed here instead of the bullock as in the northern States. Neither is the elephant here domesticated, so that Pahang belongs, in these respects, rather to the southern than to the northern section of the Peninsula. The only other vegetable products are jungle produce and some timber, which has of late years been exported mainly by the "Pahang road" to Kuàla Lumpor.

The capital of the State is *Pekan*, which lies a few miles from the mouth of the Pahang. Other settlements are *Chĕno*, some way up the main river; *Temerloh*, near the river Semantan; *Tanjong Besar* and *Penjom* on the river Lipis; *Jĕlei*, centre of the gold industry; and *Temelin*, noted for its earthenware.

The inland communications are chiefly by means of the wide-branching river system. There are no roads, and the jungle tribes are mainly confined to the inland connections with Kelàntan and Tringgànu; a path crossing from Ulu Kwantan through Perim to Ulu Lŭet, and another from Ulu Bĕrà through Paso to Ulu Keratong. There is also a way from Geliting to the R. Lipis, used by Malays passing between Pérak and Pahang. This has been recently (1884) explored by the Pérak officials. A road has also been made by the Sĕlàngor Government, connecting the frontier at Ginting Bĕlei (2300 feet) with Kuàla Lumpor.

The Government of Pahang is practically independent and arbitrary. It has always looked to the south, formerly to Johor, and of late years to Singapore, for support and protection, especially against Siam. But the Bendahāra, who has recently assumed the title of Sultan, always exercises despotic power within his jurisdiction. The revenue appears to be small, the great national wealth of the country being entirely undeveloped. There are but few Chinese settlers, and the trade is chiefly in the hands of the Bendahāra.

The local history is obscure, and appears to have been mainly concerned with invasions and threats from Siam. To a great extent Pahang escaped the troubles from which Johor suffered through its Portuguese and Dutch neighbours. Unlike the Malay States, it has of late years been growing more rather than less independent. The present ruler, then styled Wan Ahmed, obtained possession by force in 1862, when a treaty was made with Johor under the sanction of the Straits Government. In virtue of this treaty the long-disputed boundary with Johor at the river Endau was settled in 1868 by the arbitration of the British Governor. Thus was created some dependence on the part of Pahang, and on the part of the colonial Government some obligation of protection and recognition.

4.—JOHOR.

Johor (Jehór), **which** since 1877 comprises **Moar**, includes the whole of the southern end of **the** Peninsula, from 2° 40′ N. lat. to Cape Romania, together with the adjacent islets. It is surrounded on three sides by the sea, the inland boundaries being Malacca, Jehol and the river Endau towards Pahang. Although the area must be nearly 9000 square miles, the population, chiefly agricultural, is probably little over 100,000, mostly confined to the districts lying near Singapore on the one side and Malacca on the other.

The interior, still mainly under virgin forest, **and but partially** explored, **is on the** whole less mountainous **than any** other part **of** the Peninsula. The hills form detached groups **or** portions of two interrupted chains running along the west and east sides, one from Mount Ophir by Penggālam, and Mount Formosa to Pulei and the Carimons group, the other from the Segāmat Hills and Mount Janing to the Blūmut and neighbouring heights beyond Menhābak and Panti.

The Blūmut Hills (3300 feet) are the chief mountain range, giving rise to the river Johor flowing south, to the Sedili flowing east, and to the Kahang, which runs north to the Sembrong, an affluent of the Endau. Mount Ophir **in** Moar (4050 feet) is probably the highest peak in the State, and was long regarded as the highest in the Peninsula. Its shape and position near the sea are remarkable, and although it gives rise to no large rivers, two of its streams, **the** Chohong and Cemas, have some importance as forming **the northern** frontiers of **Johor** towards Malacca and Johol. Ophir was **so named** by the earlier European explorers everywhere in **search of the region** whence Solomon obtained his gold.

There are three considerable rivers in Johor, the Endau, Johor, and Moar, of which the last-named is the largest and most important in the southern extremity of the Peninsula. It takes its rise in the Négri Sémbilan territory, and after collecting numerous waters from the inland uplands flows from Brimbûn (Berembûn), south-westwards to the west coast below Malacca. The Endau, which forms the boundary towards Pahang, runs from the Segamat Hills in a north-easterly direction to the east coast, while the Johor flows from Mount Blûmat southwards to a wide estuary opposite Singapore.

The majority of the inhabitants of Johor are Chinese, who are concentrated, as cultivators of gambier and pepper, mainly in the extreme **south** over against Singapore, of which Johor has been called the "back country." From Singapore they cross over to the mainland; the capitalists for whom they work are Singapore traders, and their produce, with most of their earnings, finds its way **back to the** same place. Of late years European speculators have begun to plant sago, tobacco, coffee, tea, and cocoa, on a large scale in Bâtu Pahat, and some other districts. The collection of gutta for **the** Singapore market, after the discovery of its useful properties by Dr. Montgomerie in 1842, was carried on actively till the supply was exhausted. Saw-mills have also been worked with some success; but at present the chief exports are gambier, pepper, tapioca, timber, rattans, and damar, for which Singapore is the chief outport.

The only mineral found in abundance is iron, which, although occurring almost everywhere, is nowhere worked. Gold is known to exist in one or two places, and tin in several districts, but no tin-mining is actively carried on except in the Carimons Islands, which belong geologically to Johor, although now politically separated from that State, **and under** the Dutch flag.

The present capital is *Johor Bâru*, or *New Johor*, which has taken the **place** of *Johor Lâma*, or *Old Johor*, situated a few miles up the Johor estuary. The new town is a flourishing little place, lying about 14 miles north-east of Singapore city, in 1° 26' N. lat. There are no other towns properly so called; but on the south bank of the Moar stand the populous villages of *Lenya, Bandar Mâharâni*, and *Bukit Kepong*. Further south is the populous Javanese settlement of Padang, which, unlike most places in the Peninsula, lies not on a river, but on the sea-shore, which is here open and sandy.

Although the Government is of the usual autocratic character, the freedom and enlightenment of its administration contrast favourably with the systems prevailing in most other Malay States. For the last twenty-five years the country has been ruled wisely by the Mâharâja Abubâker, K.C.S.I., who visited England in 1866, 1878, and 1886, and who has done much to develop the resources of the land. His Chinese subjects, by nature indifferent to their rulers, provided their personal independence is secure, have hitherto given little trouble to the authorities, even where they are in a large majority. This is generally true of other States, with the exception of disturbances in certain mining districts, such as the troubles at Lūkut in 1834, and Lārut in 1872.

Johor, whose history extends back to the Portuguese days, took an important part in the 140 years' struggle over Malacca between the Portuguese and the Dutch. At the beginning of this century, the central authority having been removed from the mainland to the Lingga and Riau archipelagos, little cohesion remained among the different feudatories, and the hereditary princes of Pahang and Bûlang became virtually independent. At the restoration of the Dutch possessions at the peace, all the former dependencies of Johor, including Bûlang and the Carimons, were unfortunately assigned to Holland, the Johor rule being thus henceforth confined to the mainland and contiguous islets.

Since then the principal changes have been those resulting from the establishment of Singapore; from the Treaty of 1855 with the British Government, by which the Temenggong's *de facto* administrative rights in Johor were acknowledged; and lastly, from the restoration of Muar to Johor in 1877. Since 1868 the ruler has enjoyed the title of Mahārāja, not previously known in Malaya.

5.—THE STRAITS SETTLEMENTS.

The **Colony** of the Straits Settlements, which comprises Singapore, Penang (with Province Wellesley and the Dindings), and Malacca, now contains about 1500 square miles, and nearly 500,000 inhabitants. The settlements were transferred from the control of the Indian Government to that of the Secretary of State for the Colonies **on** the 1st April, 1867, by an Order in Council issued **under** 29 & 30 Vict. c. 115.

The earliest settlement was Penang, which was founded in 1786. Its foundation is something more than the commencement of the Colony, for it marks the beginning of the **enormous** trade, and was in some sense the forerunner of all the colonising enterprise, in the parts beyond India—Malaya, China, and Australasia. It may be noticed that, within a few months of the time Captain Light first anchored in Penang harbour, the earliest expedition to Botany Bay arrived at Port Jackson. When in 1796 Penang became the Penal Station for India, there was some superficial resemblance between the two infant settlements and the two enterprises, which have both developed so enormously during the present century. The immediate prosperity of Penang, and its superiority to the Company's trading station at Bencoolen, attracted Chinese traders, and still more Chinese settlers, and gave an early impulse to **the** expansion of its **commerce. The** troubled times of the **great** European war, which commenced seven years **after** the foundation of Penang, brought special opportunities to **this** outlying station; and at the close of a single generation the little Settlement had become a power in Malaya, under the direct and indirect influence of which the " Dutch monopoly system " had been completely overthrown. The British possession of the Straits, after 1795, became secure, first through our holding Malacca, and when that was given back **by** the establishment of Singapore.

The settlements were not formed into one Government till 1826. But the Straits have, since 1795, been, in every sense, a British possession, our power being paramount on the **western** or navigable shore.

The colony has hitherto been little more than a place of trade; **and** though it is now beginning to show some development in other directions, yet, from its circumstances, trade must always be its principal feature. As a market alone, it ranks, with Hongkong **and** Malta, not only above all other Crown colonies, but with a gross total of imports and exports which, excluding those two trade centres only, exceeds that **of all** other such colonies put together. For 1884 its total trade was at the extraordinary rate of above £85 a head of the population, a rate exceeding that of the United Kingdom **and** its most prosperous colonies in Australia, and **prob**-ably of any other country in the world.

The early prosperity of the colony resulted from **its central** position as a port of call for European, Indian, and Chinese trade. The local trade, for which both Singapore and Penang are so well placed, and which now forms so much **more secure** a basis of future prosperity, has taken time to develope. But within the last few years it has rapidly assumed increased proportions, and already far exceeds the ocean-going trade.

At the Transfer, the United Kingdom trade with the colony was £3,476,000, and the local trade (including Netherlands India and **the** Malay peninsula), £2,669,000; but now the position is reversed; **that** of the United Kingdom for 1885 amounting to only £5,500,000, while there is a local trade of nearly £13,000,000.

A similar change has been in progress, **on** a smaller **scale, in** the trade with India, as compared with the essentially local **trade** with the Malay Peninsula. With these facts established, there can be little to fear from any change in ocean routes. The colony will find its surest guarantee of continuing prosperity in the growing propor-**tions of** the trade done with its immediate neighbours, for which its **situation** makes it the natural metropolis.

The population of the colony was, according to the census of 1881, 423,384 ; in 1856 it was 248,000, and will thus have doubled itself in a generation. The Chinese and Indian population have greatly increased, but can only be maintained at their present figures by immigration, for the women number but a fourth of the men. Among the Malays, the sexes are almost equal in number ; and the increment, which in their case amounts to 2 per cent. per annum, is a natural increase, due to a high birth-rate, and not dependent on immigration.

Vegetable Products.—The flora of the colony is very rich in variety of forms. The number of flowering plants has been estimated at about 5000, and the flowerless kinds at about 300 ; but a great number of the flowering kinds produce inconspicuous blossoms, and so are commonly supposed to be without flowers. The trees producing valuable timber may be put at 100 kinds, of which the best are the talau, tampinis, seraya, meranti, daru, kladang, kûlim, petaling, rengas, merbau.

Of native fruits there are about nine varieties in daily use, supplemented by about six introduced kinds, including the pineapple and orange. The culinary vegetables are chiefly acclimatised Chinese kinds, comprising lettuces, beans, radishes, &c. of a much inferior sort to the similar European vegetables. The vegetable products which form part of the exports of the colony are about 40 in number, of which pepper, sugar, tapioca, indigo, coffee, cocoa-nuts, sago, gutta-percha, caoutchouc, and canes are the principal. The well-known Malacca cane is not, however, found in Malacca, but only in Sumatra and Borneo.

Gutta percha (*Gîtah*) deserves special mention. The plants that produce it, of a commercial standing, are about 20 in number ; about ten of which are trees, and 10 creepers, *Gîtah Tabon*, the produce of a tree, being the best known.

The Straits sago is chiefly produced by a large palm which grows in swampy places, from the pith of which sago is made. The kinds of oil exported are five in number, among which an essential oil, extracted from the lemon-grass, is the most important. Tea, coffee, and chocolate are not yet produced in any large quantity, but Liberian coffee promises to do well. Among spices, nutmegs, cloves, pepper, and cinnamon are exported, the pepper in large quantities, though most of it is not grown in the colony.

The grape-vine is not found native in the colony, and only succeeds with great difficulty under cultivation. Native vines with clusters rivaling those of the grape-vine in beauty, but uneatable, are, however, found in great plenty.

Of late years, both public and private enterprise have been active in introducing various foreign plants which yield valuable foreign products : among more recent ones may be enumerated the teak tree of India, the Brazil-nut tree, and American and African india-rubber-producing trees. The Queensland nut-bush and numerous other useful and European plants are being tried on the hills, with more or less success.

A curious feature of the vegetation of the colony is the appearance of many Australian plants on the higher hill-tops. The beautiful Victoria regia lily of the Amazons grows well, and many other introduced plants have become acclimatised in gardens and by the way-side ; but owing to the stimulating nature of the climate, few of them produce flowers or fruit as freely as in their native *habitat*, while leaves and branches flourish much more freely.

Many products, **once abundant in the colony, have** become comparatively rare, through wasteful habits and the want of any systematic conservation ; in fact many have retired considerably beyond the limits of the Settlements, and the Government of the colony has taken steps to re-establish some of these by growing young plants on waste-lands and in forest reserves.

Minerals.—No minerals are found in any workable quantities, except a little tin **in** the south of Malacca. This is natural from the situation of the Settlements, lying as they do upon the coast of the Straits. Almost immediately beyond the frontier it happens that both in Province Welles-ley and Malacca valuable tin deposits have been worked, and in the latter Settlement some gold-diggings also at Chinderas, near Mount Ophir. The Palæozoic rock occurring so frequently throughout the Settlements is largely charged with iron ores, which under the action of weathering are converted into a red limonite or laterite, forming **a** most durable building material.

Government.—The Government is of the usual type **in** British Crown colonies. It is ranked as a "first class" colony, *i. e.* the Governor's salary comes within the category " £5000 and upwards." The Governor has also general control over the Protected Native States.

The colony's revenue is now about £720,000 a year, and a municipal revenue of about £106,000 more is collected separately. The **rate** con-tributed is thus nearly £2 a head of the population, which, though lower than the rate in the Australian colonies, stands highest among the Crown colonies.

SINGAPORE.

SINGAPORE **is** an island about 27 miles long by 14 wide, containing an area of 206, or, with the adjacent islets, 223 square miles, situated at the southern extremity of the Malay Peninsula, within 80 miles of the equator. Time, 6 hours, 55 minutes before that of Greenwich. The island is separated from **the** continent by a narrow strait (*Sélat Tébrau*) about three-quarters **of** a mile in width. All the small islands within ten miles of **its** shores form part of the Settlement.

The seat of Government, for the whole colony as well as the Settlement, is the town of Singapore, at the south of the island, in lat. 1° 17′ N., and long. 103° 50′ E.

Singapore was occupied by Sir Stamford Raffles, with **the** consent **of** the Governor-General, **in** February 1819, under an agreement with the Princes of Johor. In 1823, it was transferred to the direct Government of Bengal, and in 1826, incorporated with Penang and Malacca, and placed under the Governor and Council of the incorporated Settlement. It became the recognised seat of Government in 1837.

The surface **of the** island is undulating, nowhere over 500 feet **high, and** consisting **of** laterite **resting on** sandstone. Granite is **found in** a few places, principally **to the** north and east. Gambier,

44 EASTERN GEOGRAPHY.

indigo, pepper, and many local fruits and **vegetables** grow well; but the Settlement depends for rice on the neighbouring countries of Java, Camboja, Burma, and Bengal.

PENANG.

PENANG is the name both of an island, and of the Settlement **for** which it is the seat of local administration.

The Settlement has altogether an area of about 600 square miles. The island, officially called Prince of Wales' Island, is about 15 miles **long** and 9 broad, containing an area of 107 square miles, situated **off** the west coast of **the** Malay Peninsula in 5° N. latitude, **and** at the northern end of **the** Strait of Malacca.

On the opposite mainland, from which the island **is** separated by a channel a few miles broad, lies Province Wellesley, a strip of territory containing 270 square miles, and forming part of the Settlement. The province averages 7 miles in width, and extends 45 miles along the coast; it includes, since the Pangkor Treaty (1874), about 25 square miles of newly-acquired territory **to** the south of the Krian. About 200 square miles of land in the Pangkor Islands and opposite coast are also comprised in this territory, and form the so-called Dinding Settlements.

The chief town is George Town, in 5° 24' N. lat. and 100° 21' E. long. The local government of the whole Settlement **is** administered by **a** Resident Councillor.

MALACCA.

MALACCA is situated about one-third of the way up the western coast of the Peninsula, between Singapore and Penang, about 110 miles from the former and 240 from the latter, and consists of a strip of territory 42 miles in length, and from 8 to 25 miles in breadth, containing **an** area of 659 square miles.

Tin is found at the Linggi sands and **a few other** places, and gold **on the** slopes of Mount Ophir. Hot springs, noteworthy as being **the** only sign of volcanic agency in the Peninsula, occur at Ayer **Panas** and near Pulan Sebang, some 20 miles from Malacca. They are said to possess salubrious qualities.

The principal town, called Malacca, is in 2° 10' N. lat. and 102° 14' E. long. The local government is administered **by a** Resident Councillor.

STATISTICS.

AREAS AND POPULATIONS.

	Area in Sq. Miles.	Population.
Malay Peninsula .	75,000	1,200,000 (?)
Siamese Malaya .	40,000	400,000
British Malaya . .	35,000	800,000

Siamese Division:

Ligor and Senggora .	17,000	150,000 (?)
Kedah . .	4,000	120,000 (?)
Patâni States .	6,000	50,000 to 75,000
Kêlantan . .	7,000	200,000 (?)
Tringgânu with } Kemâman } . .	4,000	20,000

British Division:

Pêrak . .	8,000	118,000 (1884)
Sêlângor with Klang	5,000	50,000 (1884)
Sunjei Ujong .	500	14,000
Nêgri Sêmbilan .	2,000	30,000
Pahang .	10,000	63,000
Johor .	8,500	100,000
Straits Settlements .	1,460	over 500,000 (1886)

POPULATION OF THE PENINSULA ACCORDING TO RACE.

Siamese Division:

Siamese	100,000
Sam-Sams	50,000
Malays	150,000
Negrito Wild Tribes	10,000
Chinese	90,000
	400,000

British Division:

Malays	350,000
Negrito Wild Tribes	10,000
Chinese	300,000
Klings (Indians)	40,000
Miscellaneous	100,000
	800,000

Total Malays	500,000
Total Chinese	390,000
Malays in the Straits Settlements . .	174,326 (1881)
Chinese „ „ „ „ . .	174,327 „

KĔDAH.

Population (1820)	. . . . . over	50,000
„ (1839)	. . . . .	21,000
„ (1886)	. . . . .	100,000 (?)
„ According to Carl Bock (excessive)		525,000

PATĀNI.

Population (1786)	. . . . .	115,000
„ (1832)	. . . . .	54,000
„ (1886)	. . . . 50,000 to	75,000
„ According to Carl Bock	. .	200,000

KĔLANTAN. .

Population (1856)	. . . . .	50,000
„ (1886)	. . . . .	200,000
„ According to the natives	.	600,000
„ of Kĕlantan town	. . .	20,000
Annual export of tin	. . . .	3000 pikuls

TRINGGĀNU.

Population (1856)	. . . . .	37,500
„ (1886)	. . . . .	20,000
„ Tringgānu town (1856)	. .	15,000
„ „ „ (1886)	. .	2,000

PĔRAK.

Malay population (1884)	. . . .	70,000
Chinese „ „	. . . .	48,000
Thaipeng town, pop. „	. . . .	35,000
Tin exported (1883)	. . . .	£400,000
Other exports „	. . . .	£700,000
Total imports and exports (1883)	.	£2,100,000
Revenue (1884)	. . . . .	£306,000
Expenditure (1884)	. . . .	£296,000

SĔLĀNGOR.

Malay population (1886)	. . . .	20,000
Chinese „ „	. . . .	30,000
Exports (1884)	. . . . .	£400,000
Revenue „	. . . . .	£89,000

SUNGEI UJONG.

Population (1886)	. . . .	14,000
Revenue (1884)	. . . . .	£24,000
Expenditure	. . . . .	£28,000

NÉGRI SÉMBÎLAN STATES.

Jëlëbu, population	700
Johol ,, ...	5,000
Sri Ménanti	3,000
Rémbau	12,000

PAHANG.

Malay population (1886)	60,000
Sakei (Negritoes) ,, ...	3,000
Chinese ,, ...	500

JOHOR.

Malay population (1886)	40,000 (?)
Chinese ,, ,, ...	60,000 (?)
New Johor ,, ,, ...	15,000
Padang ,, ,, ...	2,000

STRAITS SETTLEMENTS.

Pop. 1881.

SINGAPORE :

Chinese	86,766
Malays	22,114
Klings (Indians)	12,104
Europeans	2,769
Miscellaneous	15,455
Total	139,208

PENANG :

Chinese	67,820
Malays	84,724
Klings	27,115
Europeans	674
Miscellaneous	10,264
Total	190,597

MALACCA :

Chinese :	19,741
Malays	67,488
Klings	1,887
Europeans	40
Miscellaneous	4,423
Total	93,579

Total pop. Straits Settlements (Census 1881)	423,384
,, ,, ,, estimated 1886 over	500,000

Areas.

						Sq. Miles.
Singapore	.	.	.	.	.	. 206
Penang with Dindings	.	.	.	.	.	. 580
Malacca	.	.	.	.	.	. 659
				Total.		1445

Trade and Finance.

Total trade with Gt. Britain (1868)		.	£3,476,000
„ local trade		„ . .	2,669,000
„ „	(1884) nearly		13,000,000
„ trade with Gt. Britain	„	. .	5,500,000
„ „ India (1868)		. .	1,968,000
„ „ „ (1884)		. .	1,500,000
„ „ Malay Peninsula (1868)			839,000
„ „ „ „ (1884)			2,200,000

Exports to Gt. Britain.		*Imports from Gt. Britain.*
1880.	£3,698,000	£2,269,000
1884.	4,612,000	2,633,000

		Singapore.	Penang.	Malacca.	Total.
Imports.	{ 1880.	£10,872,000	3,534,000	644,000	15,050,000
	{ 1884.	14,257,000	5,616,000	532,000	20,405,000
Exports.	{ 1880.	9,779,000	3,554,000	651,000	13,984,000
	{ 1884.	11,675,000	6,450,000	548,000	18,673,000

Shipping (1884).

Foreign vessels entered	5,848	Tonnage	3,634,000
Native „ „	9,417	„	266,000
Foreign „ cleared	5,759	„	3,576,000
Native „ „	9,849	„	281,000

	Revenue.	*Expenditure.*
1880.	£327,000	£335,000
1885.	654,000	648,000

Public debt (1886), £56,000.

PART II.

INDO-CHINA.

CHAPTER I.

Position—Boundaries—Extent.—The term *Indo-China*, for which alternative expressions are *Further India* and *Transgangetic India*, was originally proposed by Malte Brun for the easternmost of the three great Asiatic peninsulas, forming the south-eastern limb of the continent. It thus corresponds in position and some other respects with the Balkan Peninsula of the European Continent, and, like it, is continued south and south-eastwards by numerous insular groups, through which it gradually merges in the Australian mainland.

Washed on the west by the Bay of Bengal, which here develops the Gulf of Martaban, on the south and east by the China Sea, with the corresponding Gulfs of Siam and Tonkin, Indo-China abuts on its north-west frontier with India, and on the north with China. But the term Indo-China was suggested not so much by this geographical position, as by the twofold origin of its religious and social culture, derived partly from China, but to a much greater extent from India. Hence the alternative expression *Transgangetic India*, strictly correct in a geographical sense, may also be justified on historic grounds.

Excluding from consideration its extreme southern prolongation through the Malay Peninsula, which is treated separately in this series, Indo-China presents a somewhat compact oval form disposed in the direction from north-west to south-east, and comprised almost entirely between 10° N. lat. and the Tropic of Cancer, but projecting in the extreme south-east to Cape Camboja (9° N. lat.), and in the extreme north-west to about 27° N. lat. to the Patkoi Mountains here

E

separating it from Assam. The longitudes of 92° and 109° E. mark
its extreme western and eastern limits on the Bay of Bengal and
China Sea, the total length in the direction from north to south
being about 800 miles, from the Chinese frontier to the Isthmus
of Kra, and 950 at its broadest part, between the Ganges delta and
Gulf of Tonkin, with an approximate area of about 800,000 square
miles.

Mountain Systems.—The **salient** physical features of this
region present a certain simplicity of outline, as shown especially
in the uniform and nearly parallel disposition of its mountain
ranges and river valleys, which run mainly in the direction of its
long axis from north-west to south-east. Projecting like Southern
China, with which it forms a geographical unit, from the elevated
Tibetan plateau, the peninsula is of an extremely rugged character
in its northern section, where it begins to fall rapidly towards the
central plains. Here the slope of the land is clearly indicated by
the numerous falls and rapids obstructing the upper courses of the
main streams and their chief affluents. But the mountain chains
forming the water-partings between these river basins, although
seldom exceeding 7000 or 8000 feet, maintain a mean elevation **of**
from 5000 to 6000 feet throughout their whole course to the Malay
Peninsula.

In the north-west the ARAKAN YOMA range, separating the Arakan
coastlands from Upper Burmah, has several crests from 6000 to 7000 feet
high, culminating in the Malselai Mon or "Blue Mountain" (7100 feet)
in the Lushai country. This range, which terminates at Cape Negrais at
the western angle of the Irawaddy delta, is crossed by several passes, of
which one of the most frequented is that of Ayeng (under 4000 feet)
leading from the coast to Upper Burmah. The prevailing formations are
limestones and sandstones of the chalk and tertiary periods, interspersed
with some eruptive rocks, but no active volcanoes. But on the coast and
adjacent archipelago are grouped a large number of mud volcanoes, as
many as fifteen in the island of Ramri alone, all subject to frequent and
violent eruptions.

Parallel and east of the Arakan Yoma is the PEGU YOMA range,
forming the divide between the Irawaddy and Sittang basins, but seldom
exceeding 2000 feet. Southwards it merges in the extensive plain of
Pegu, formed by the united lower valleys and deltas of the Irawaddy and
Sittang, and stretching from Cape Negrais to Martaban west and east.
The Pegu Yoma, one of whose crests, the Puppa or Pappa Dúng (3000
feet), presents the character of an extinct volcano, is continued northwards
by the SHAN-YOMA, which separates the waters flowing west to the
Irawaddy and Sittang basins from those draining east to the Salwin. This
range, which rises in the north to over 10,000 feet, and even in the south
has one peak, the Nattung, 8000 feet high, appears to consist mainly of
slaty clays alternating with sandstone, and here and there assuming a
basaltic character. Stratified sandstones interspersed with veins of quartz
are also a prevailing feature of the Tenasserim hills, which form a southern

prolongation of the northern "Yomas" or "Mountains," and which in some places attain an elevation of 5000 or 6000 feet. Beyond Tenasserim the system is continued through the Malay Peninsula as far as the island of Billiton below the equator.

The general geological structure of Burmah is described by M. R. Romanis as very simple, the chief formations running north and south in great mountain ranges. The tertiary formations of Pegu reach northwards to the great bend of the Irawaddy at Kyônta-lung below Ava, while the metamorphic rocks of the Martaban hills are continued in the Shan Yoma east of Mandalay. In the same way the limestone of the Salwin and Kachin hills corresponds with that of the Dawna range east of Maulmein, and the general features of the country much resemble those of the north-west provinces of India.

East of the Burmese and western Siamese ranges the orographic system becomes more irregular and less elevated, the central Siamese plains being broken only by low and short ridges or isolated eminences, such as the Koh Sarap east of Shantabûn (2100 feet), the Prabat and Bassac hills (3800 feet). But the regular and parallel disposition of the Indo-Chinese ranges reappears further north and east, in the chain separating the Mekhong from the Song-koi basin, and in the Cochin-Chinese coast range. North-west of Hué the main range rises to a height of from 6000 to 7000 feet. Further south is developed to lofty SAKAVAN or BOLOVEN PLATEAU (about 3000 feet), which is enclosed between the Mekhong and Don rivers on the west and north-west, and by the Kong on the east and south-east. This extensive tableland, which has been explored by Thorel, Harmand, and other French naturalists, presents in some places the aspect of a grassy or swampy steppe, and in others is covered with dense forests of conifers, oaks, chestnuts, intermingled with palms, bamboos, and other subtropical species. The soil consists of a ferruginous clay resting on sandstones interspersed with lavas and scoriæ, which combined with the presence of hot springs and several cone-shaped crests, show that this region was formerly the scene of igneous activity.

In the extreme north-east the hills and plateaux merge everywhere in the low-lying plains of Camboja, which are interrupted in the east by the granitic TIONLAT (over 3000 feet) about the sources of the Donnai, in the west by the Prabat and PURSAT HILLS between the Tonlé-Sap basin and the Gulf of Siam, culminating in the Elephant Peak (3000 feet) west of Kampot.

Mineral Wealth.—The Shan uplands and the ranges separating the Irawaddy and Salwin basins contain rich iron, lead, copper, tin, and silver deposits. The Shurili river washes down golden sands from Yunnan; rubies (spinels?), sapphires, emeralds, topazes, and other precious stones were for ages collected in the hills to the north-east of Ava for the royal treasury of Burnah. This is also one of the few regions containing deposits of jade, which occurs chiefly in the Mogûng district north of Bhamô. Saline springs and petroleum are found in great abundance at the eastern foot of the Arakan Yoma, where over 500 wells have been sunk near Yay-nan-gyung, on the left bank of the Irawaddy. The yield now exceeds 12,000 tons yearly, some of which is exported to Great Britain.

E 2

"This petroleum was one of the royal monopolies, and large quantities used to be shipped to Rangûn to be manufactured into pagoda candles; but the American rock-oil and the development of the Baku wells (in Caucasia) interfered greatly with the sale."— (*J. G. Scott*, p. 59.) Coal, slaty and bituminous, occurs both in Tenasserim, where it has never been worked, and on the Irawaddy, where it has been long worked by the Burmese, especially at Thinkadaw, some 30 miles above Myin-gyan, at Shin-pagah, midway between Mandalay and Bhamô, and in the Shan hills east of Mandalay. From Payen-tûng, 150 miles north of Bhamô, come large quantities of amber, which is much used for ear-plugs, necklaces, Buddhist rosaries, and similar objects by the natives. Platinum also is said to occur **near** Kanni on the Chindwin river, and iron and silver mines were **once** largely worked, but are now abandoned.

In Siam occur rich deposits of copper, tin, antimony, and magnetic iron, and in the mountain range between the Mekhong and Red River iron, tin, copper, silver, and gold. Near the delta of the latter river the French engineer, M. Tucks, discovered in 1881 an extensive coalfield over 40 square miles in extent. It also seems probable that the valleys of both head streams of the Song-koi (Red and Black Rivers) abound in all manner of mineral ores, which, however, cannot at present be utilised, owing to the unsettled and inaccessible character of the country.

River Systems.—The disposition of the river systems is marked **by even** greater **uniformity** than that of the **mountain** ranges. Excluding the smaller streams flowing independently to **the** coast in Arakan, Tenasserim, and Annam, the whole of the peninsula drains to the surrounding waters through five great fluvial basins, all pursuing a more or less parallel southern or south-eastern seaward course. Thus going eastward, the IRAWADDY and SITTANG collect in a common delta nearly all the waters of Upper and Lower Burmah; the SALWIN **takes** up all the drainage of north-east Burmah and the borderlands between Lower Burmah and west **Siam**; through the MENAM all the streams of Central Siam find their way to the Gulf of Siam; the whole of the Lao country (North and East Siam), Camboja, and Lower Cochin China are comprised in the basin of the MEKHONG (CAMBOJA) river; lastly, the SONG-KOI (RED RIVER), with its many-branching delta, carries to the China Sea all the surface waters of Tonkin.

Two only of these basins, the Irawaddy-Sittang and the **Menam,** belong entirely to the peninsula. Of the others the Salwin **and** Mekhong have their source on the Tibetan plateau far away **to the** north-west, while the Song-koi takes its rise on the **rugged** Yunnan tableland in south-west China.

Above Bhamô the *Irawaddy* forks off into two head-streams, the Myit Guey ("Little River") to the east, and the Myit Gyi ("Great River") to the west, both flowing from the still unexplored region about the Tibeto-Chinese frontier. But so great is the volume of water at Bhamô (24° N. lat.) that the western branch was long supposed by Elisée Reclus, Robert Gordon, and others to be identical with the great Sanpo river of Tibet. This important hydrographic question, however, has at last been settled by the Indian pandit known as A. K., who has shown that the Sanpo cannot possibly flow to the Irawaddy, and Mr. J. F. Needham's still more recent exploration (1885) has identified it beyond all doubt with the Dihong (Brahmaputra). At the junction of the two forks the Irawaddy is already 900 yards across, so that one or other of them, as suggested by J. G. Scott, may perhaps have its source in some large inland lake. In any case the Irawaddy is one of the finest streams in the world, navigable for steamers of considerable size from its delta in the Gulf of Martaban for over 800 miles to Bhamô, near the Chinese frontier. But about 20 miles higher up it passes through a dangerous defile, where the stream, suddenly narrowing from 1000 to 150 yards, rushes with great velocity between sheer rocky walls, and where the whirlpools and backwaters render all navigation impossible except for very small craft. The Irawaddy has a probable length of 1000 miles, with a mean discharge of 480,000 cubic feet per second in the delta, rising during the floods to nearly 2,000,000 cubic feet.

The *Sittang*, which drains an area of over 20,000 square miles in the quadrangular district formed by the Pegu Yoma and Punglung Hills, belongs, strictly speaking, to the Irawaddy basin. It flows in exactly the same direction as the main stream between Bhamô and Mandalay, and after a course of some 330 miles unites with it in a common delta. During the rainy season this low-lying watery region presents an intricate maze of channels and backwaters stretching round the Gulf of Martaban from Cape Negrais to Maulmain, and here intermingling with the waters of the Salwin.

Although containing a much smaller volume, the *Salwin* has a far longer course than the Irawaddy. It has been clearly identified with the *Lu-Kiang* (*Lutze-Kiang*), which rises on the Tibetan plateau, probably about 34° N. lat. 92° E. long., flowing thence for hundreds of miles in its deep and narrow rocky valley between the Irawaddy and Mekhong basins first south-east, then due south through eastern Burmah to its mouth at Maulmain on the Gulf of Martaban. Below the Thung-Yang confluence its lower course is obstructed by dangerous rapids, practically barring all navigation for the greater part of the year. Hence, notwithstanding its great length and depth, the Salwin is of little use as a water highway. It has a mean discharge of from 300,000 to 400,000 cubic feet per second.

The *Menam*, or "Mother of Waters," stands in the same important relation to Siam that the Irawaddy does to Burmah. Throughout the greater part of its course from the Lao uplands to its mouth at the head of the Gulf of Siam it is navigable for small craft, while steamers ascend the main channel with the tide as far as Bangkok. During the rains the Menam floods its banks for miles in all directions, ever depositing fresh alluvial soil, irrigating the rich paddy fields on the surrounding plains, and affording a large navigable area for native craft throughout the flooded tracts. The sedimentary matter thus wasted down has already advanced the shore-line many miles seawards, and is still continually encroaching on the gulf, where the deep water is separated from the plains of Bangkok by extensive mud-banks stretching for 60 miles east and west, and accessible to large vessels only at high water.

By far the longest river in the peninsula is the *Mekhong*, familiarly known as the *Camboja*, which under the name of the *Lantzan-Kiang*, or *Kinlong-Kiang*, has its source on the Tibetan tableland about 34° N. lat., 94° E. long. Throughout its upper course it flows in a narrow, deep defile between the Salwin and Yangtse-Kiang, through an unexplored region inhabited by the Lyssu, Mosso, and other semi-civilised peoples on the borderland between Tibet and China. Below the confluence of its great affluent, the Semun from the west, the lower course of the Mekhong is obstructed by the Khong rapids, which are scarcely surpassed in extent by those of any other river in the world.

At Pnom-penh about the head of the delta, some 180 miles from the sea, it receives the overflow of the Tonlé-sap, an extensive sweet-water reservoir commonly known as the "Great Lake of Camboja." During the floods between June and October this lacustrine basin is nearly 70 miles long by 15 broad, with a depth of over 40 feet, and an area of about 1000 square miles. At this time the lake is fed by a backwater from the Mekhong; but at low water the current is reversed, and the lake discharges into the river. It teems with fish, of which about 10,000 tons are annually cured and exported to the surrounding lands.

Below Pnom-penh the main stream ramifies into two channels, the Han-giang, or "River of Bassac," in the west, and the Tien-giang in the east, which flow in a nearly parallel course for about 120 **miles** through the delta. The eastern river develops numerous secondary branches, of which the most important are the Donnai (Dong-nai) and the river of Saigon ; from the western river several channels are also thrown off, some of which now flow west to the Gulf of Siam. Thus the greater part of Lower Cochin-China belongs to the Mekhong delta, which has a coast-line of 380 miles, besides shallows and sandbanks stretching for some 30 miles seawards. It has a mean discharge of 420,000 cubic feet per second, falling at low water to 50,000, and rising during the floods to upwards of 2,300,000.

The basin of the *Song-koi* (Song-kai, Song-tha), properly *Shong-kai*, that is, the "Great River," the "Red River" of European writers, comprises, with its two chief tributaries, the Song-bo or "Black River, and the Mi-lei-ho, nearly the whole of Tonkin, and a considerable part of south-east Yunnan. On the Chinese frontier, 300 miles from the coast, it is already about 1000 yards wide and navigable for boats ; but both the main stream and the Song-bo, which joins it below Hung-hao, also from Yunnan, are much obstructed by rapids. M. d'Augis, who ascended the Black River in 1881, counted fifty-four rapids up to Wan-Gioua, and at Thae-Keu found all further canoe navigation arrested by a chaos of rocks and debris rising 23 feet above the current. Thus the expectations of the French to reach Yunnan and establish trading relations with Southern China through this artery have been doomed to disappointment.

Some 90 miles from the coast the Red River throws off the two main channels of the delta, both of which again branch off into a vast system of intricate streams, backwaters, and artificial canals, continually shifting their beds. From the northern arm, which retains the name of Song-koi, several channels flow northwards to the still more intricate delta of the Thai-binh, which descends from Lake Babé in a still unexplored frontier district. The joint delta has a coast-line of about 90 miles, and a total area of probably not less than 6000 square miles of rich alluvial land.

Seaboard.—The Indo-Chinese seaboard develops a far greater diversity of outline than any other Asiatic region, in this respect also resembling the corresponding Balkan Peninsula in south-east Europe. Below Akyab the north-west coast is fringed with several clusters of islands, including the volcanic Ramri and Cheduba, and presenting an almost fjord-like appearance. Further down the Irawaddy delta projects far seawards, terminating at Cape Negrais in the south-west, and enclosing on the east side the deep bight of the Gulf of Martaban. Below this point the coast is again broken into several little headlands, and diversified with the extensive Mergui Archipelago, which stretches for nearly four degrees of latitude (10°—14° N.) from below Tavoi to the Isthmus of Kra. On the east side Lower Siam, with Camboja and Lower Cochin-China, forms a secondary peninsula, projecting between the vast Gulf of Siam and the China Sea for over 250 miles in a south-easterly direction. Here also the coastline is broken by several conspicuous headlands, while to the Gulf of Siam on the south-west corresponds the Gulf of Tonkin in the extreme north-east. Thus the seaboard, even excluding the Malay Peninsula, has a total length of considerably over 2000 miles, which is relatively far more than that of any other maritime region in Asia. Distinct indications of upheaval have been noticed at several points, and especially along the coast of Arakan and Lower Burmah as far as the Irawaddy delta. The movement, centred about Cheduba island, is continued seawards in the Nicobar group, although the intervening Andaman Archipelago appears to be the scene of the opposite phenomenon of subsidence.

CHAPTER II.

CLIMATE—FLORA—FAUNA.

Climate.—Lying almost entirely within the northern torrid zone, Indo-China has an essentially tropical climate, with two well-defined seasons determined by the regular succession of the two monsoons. That of the south-west, prevailing from May to September, brings the moisture-bearing clouds and heavy storms from the Indian Ocean, with a rainfall of 200 inches and upwards on the Arakan coast and in the Irawaddy basin. From September to March these winds are replaced by the north-east monsoon, which is accompanied by dry weather and cool breezes. During the months of March and April, between these two seasons, the heats are most oppressive. Yet even at this period the thermometer seldom rises above 93° F. at Bangkok, usually oscillating between 82° and 86° F., and in the dry winter season falling occasionally as low as 54° at night. At Hué, in Annam, the lowest recorded has been 62°, and at Saigon 64°, the mean temperature of the latter place being as high as 80°. Tonkin and the interior of Siam are much cooler, the glass falling in both regions to 47°, and even 45° in winter. On the other hand, these countries are subject to more intense heats, so that the further we proceed north the climate becomes more continental, that is, presents greater extremes of heat and cold from season to season.

The mean rainfall, heaviest on the west coast (200 to 240 inches), diminishes to 70 at Saigon and 60 at Bangkok. A great contrast is also presented between the two slopes of the range separating the Mekhong basin from Annam and Tonkin. The east side, exposed to the drier north-east monsoon, is in many places arid and bare of timber, while on the opposite slope a rich vegetation is supported by the moist south-western trade winds.

Flora.—On the whole the Indo-Chinese flora corresponds with that of India proper. The forests and jungle of Burmah present the same variety of plants, and yield for human food and industry nearly the same cereals, fibres, gums, and other essences. Here the chief cultivated species are rice, of which Burmah is one of the great store-houses; dani, a kind of palm which yields all the sugar required for the local consumption; maize, millet, sessamum, pulse, cotton; bananas and other fruits; tea, coffee, cinchona, and tobacco, the cultivation of which is steadily increasing. Whenever the country is opened up the Burmese forests will yield an inexhaustible supply of all kinds of useful timber to human industry. Here flourish the close-grained teak, various plants yielding gums, lacquer, and oil, and on the coastlands the magnificent *Amherstia nobilis*, with its red and golden flowers.

The Siamese flora, substantially the same as that of Burmah, also includes a considerable number of Chinese species, thus showing a gradual transition between the vegetable kingdoms of the northern and southern regions. In the eastern uplands, between the Mekhong basin and Annam Himalayan, Chinese and even Japanese varieties are found intermingled with those of the peninsula, and here are also met anemones, saxifrages, and violets like those of western Europe. The flora of Tonkin and Cochin-China is altogether extremely rich, the botanical explorations organised since the French Conquest having already discovered over 12,000 species. Pandanes and various kinds of palms fringe the low-lying coastlands, while paddy fields and garden plots cover the plains stretching inland to the foot of the hills, which are in many places clothed with dense forests of teak, ironwood, lacquer, and other gum-yielding species, ebony and the precious eaglewood, burnt only in the palaces and temples of the gods. The natives, however, cultivate little except cotton, maize, bamboo, which is turned to endless domestic and industrial uses, rice, of which there are some forty varieties, and a vine, which yields a sour wine.

Fauna.—Most of the Indian animal species reappear in the region beyond the Ganges. Here elephants are very numerous, especially in Burmah ; those of the Lao country are noted for their intelligence, and the natives everywhere display great skill in capturing and taming them. The rhinoceros also, of which three varieties are known in Burmah, is sometimes tamed, as in Assam. The tiger, which roams the Annamese forests, and reaches down to the extremity of Malaya, is seldom openly attacked, but mostly taken in snares. The Annamese fauna includes, besides the wild buffalo, the *dzin*, a species of ox perhaps allied to the Chinese *mithun*. The Burmese breed of horses is highly esteemed, while those of Cochin-China are too small and weak to serve as pack-animals. In Tonkin, Annam, and Camboja poultry, ducks, and geese are more numerous than in Europe, and every house has its pig. In Burmah rats are a great scourge, and the insect world is represented by innumerable species both here and throughout the peninsula. All the rivers and marine inlets, as well as the great Cambojan lake, teem with fish, which is a staple of food amongst all the inhabitants of Indo-China.

CHAPTER III.

INHABITANTS OF INDO - CHINA — BURMESE — TALAINGS — SIAMESE—
ANNAMESE—CAMBOJANS.

Mongolic Races.—From the anthropological standpoint the great bulk of the Indo-Chinese people belong to the Mongolic family of mankind, and more directly to the Tibeto-Chinese sub-division of that group. Amid a multiplicity of national, historical, and tribal names, a substantial unity both in the physical and linguistic types is almost everywhere apparent, and it seems evident that nearly all the present inhabitants of the peninsula came originally either from the Central Tibetan plateau, following the parallel valleys of the Irawaddy, Salwin, and Mekhong rivers, or else from China, crossing the intervening highlands by passes that have been frequented from pre-historic times. Of the three main groups the western *Burmese* still show the closest affinity to the Tibetans, especially in their speech, while the Central *Siamese* and eastern *Annamese* are more nearly related to the Chinese both in type and speech. All present the same fundamental Mongolic characteristics, shown in their low stature, ranging from 5 to 5 feet 6 inches, yellowish or yellow-brown complexion (olive, old wax, leathery or cinnamon), long black hair, round in section, thick on the head, elsewhere very scanty or altogether absent, somewhat flat features, with almond-shaped, oblique eyes, broad, short, and concave nose, rather prominent cheek-bones, small hands and feet, weak lower extremities, head mesaticephalous, that is, intermediate between the round and long forms.

Languages.—The Indo-Chinese languages belong also to the same morphological order of speech as the Chinese and Tibetan, commonly described as *monosyllabic* or *isolating*. But the former designation must be rejected, since it has recently been shown that monosyllabism is not the original condition, but the result of phonetic decay in this group. In consequence of this gradual decay words originally quite distinct in form, and composed of two or more syllables, have been reduced to words of one syllable no longer distinct in form, but in pronunciation distinguished by the different *tones* with which they are uttered. Grammatical inflection has also been mostly rejected, words being thus reduced to the condition of crude and unchangeable roots standing *isolated* from each other, and acquiring their meaning mainly from their position in the sentence. Hence a better designation for this group would be that of *isolating toned languages*. The process has been carried furthest in Chinese and Annamese, which may be taken as the typical members of the family, and which have necessarily developed the greatest number of tones, ranging from four to six, and in some dialects even to eight or ten. In this respect Siamese occupies an intermediate position between Annamito-Chinese on the one hand, and Tibeto-Burmese on the other, having preserved more grammatical inflections and developed fewer tones than the former, while the Burmese, and especially the Tibetan, have retained the greatest number of grammatical forms, and are consequently spoken with the least number of tones.

The three dominant Indo-Chinese languages have long been

cultivated—Annamese under Chinese, Siamese and Burmese under Hindu influences. Hence Annamese has borrowed a large number of Chinese words, and is written in characters derived directly from the Chinese hieroglyphic system. In the same way, most of the Siamese and Burmese loan-words are taken indirectly from Sanskrit through the Pali, the sacred language of the Buddhists, and are written with alphabets derived from the same source. Here again we see how completely appropriate to this region is the term Indo-China.

Non-Mongolic Races.—Besides the Mongolic, recent French exploration has revealed the presence of a second element, centred mainly in Camboja and the Champa country, in the extreme south-eastern corner of the peninsula. This element, represented by the old Camboja (Khmêr) stock, by the Chams (Tsiams), Kûys, Stiengs, Charays, and some other semi-civilised aboriginal tribes, is distinguished by physical characteristics approaching the Caucasic type of Western Asia and Europe. The same Caucasic type occurs amongst the Lolo, Mosso, and many other aboriginal peoples in the borderlands between China, Indo-China, and Tibet, possibly indicating **the route** followed by this stream of Caucasic migration from Central Asia **to** the south-eastern extremity of the Continent. The Kûys (Khmer-dom, or primitive Cambojans **of west** Camboja), the Charays, Stiengs, and other non-Mongolic tribes **of this** region, are described generally as above the middle size, often with wavy hair, light brown or fair complexion, and more or less regular European features, in a word, "white savages of Caucasian type" (C. E. Bøillevaux). Their *untoned* speech also is fundamentally distinct from that of the toned isolating group, in some respects betraying marked affinities with the Oceanic, or Malayo-Polynesian linguistic family. This is true especially of the dialects spoken in the uplands between Annam and the Mekhong basin by the Chams (Tsiams), Sheveas, Charays, Radêhs, and other tribes whom some writers regard as scattered fragments of the Champa State, which formerly comprised most of Cochin-China and the Mekhong delta region. But whether these peoples represent a comparatively recent immigration of true Malays from Malaysia, or the original stock whence the Malays passed from the mainland to the Eastern Archipelago, is **a** question which cannot here be discussed.

In the subjoined table are comprised all the Indo-Chinese races with their chief sub-divisions.

Mongolic Stock of Toned Speech.

TIBETO-BURMESE GROUP.

RAKHAINGTHA (*Tungtha*) "Highlanders": Lushai, Shendu, Mro,
(ARAKANESE)) or *Yoma*) Khyong. Kuki, Kumi.
{ *Kyungtha* "Lowlanders") Mag, Chakma, &c., in
or "River People") the Arakan lowlands.

BURMESE PROPER: Upper Burmah and Lower Burmah.

TALAINGS or MONS: Lower Burmah (Pegu), now mostly assimilated
to the Burmese in speech and type.

SINGPHO   { the *Kakhyen* of the Burmese; N. Burmah south to
(CHINGPAU)   { Tagoung midway between Bhamô and Mandalay.
              { *Khyen* and *Yau* 20° to 24° N. lat.

KAREN { *Sgaw*) Highlands between the Irawaddy, Sittang, and
      { *Pwo*) Salwin basins, from the latitude of Mandalay
      { *Bghai*) to Tenasserim, but chiefly in the Sittang-Salwin,
                  water-parting.

THAI-SHAN GROUP.

SHAN { *Sino-Shan*: Chinese or Northern Shans, throughout S.
     {   Yunnan and thence to Bhamô.
     { *Ngiou*:   Southern Shans, Kiang-hung, Kiang-tung, and
     {   other districts N. E. Burmah and N. W. Siam.

LAO { *Lau-pang-kak* } N. and E. provinces of Siam between the
    { *Lau-pang-dun* } Siamese and Shans.

SIAMESE PROPER: Menam basin and Malay Peninsula south to 7°
N. lat.

PALUNG: Muang-lem and Maing-Kaing districts, and **S.W. Yunnan.**

KHANTI: Upper Irawaddy and borderlands **towards Assam.**

GIAO-SHI GROUP.

ANNAMESE: Dominant race in Tonkin and Cochin-China.

MUONG: Neutral zone **between** Tonkin and Yunnan.

Mongolic and Non-Mongolic Stock of Untoned Speech.

KHMÊR-DOM (*Mahai*)
(Kuy or old { *Mânh* } Compong-Soay (Camboja proper); **provs.**
Khmêr)   { *Porrh* } Mulu-Prey and Tonlé-Repan **(Siamese**
            (*Damrey*) Camboja).

KHMÊR PROPER } The dominant **race** in the kingdom of
(Cambojans) } Camboja.

CHAM PROPER }
STIENG } South Cochin China, East Camboja, and **intervening**
BANNAR } uplands.
CHABAY }

However interesting to the student of ethnology, none **of these**
numerous peoples possess much historic importance **or political**
influence except the three leading Mongolic **races—the Burmese of**
the Irawaddy basin, the Siamese of the Menam **basin, and the**
Annamese of the Song-koi basin and Cochin China.

THE BURMESE AND TALAINGS.

The Burmese, all of whom since the close of the year 1885 have
become British subjects, betray a curious combination of qualities,
in which, however, the good seems greatly to outweigh the evil
elements. Mr. J. G. Scott, who knows them well, declares that
"their very faults lean to virtue's side," and their general indolence,
overweening national vanity, and extreme sensitiveness to real or
imaginary slights, are certainly more than balanced by a remarkably
genial, cheerful, and kindly disposition. These characteristics are
shown not only in their somewhat excessive love of pleasure, but
also in their friendliness and hospitality towards strangers, in their
boundless liberality to their Buddhist priests and teachers, and
especially in the gentle treatment of their women, who enjoy a
degree of personal freedom scarcely to be elsewhere paralleled
amongst Asiatic peoples. Like all genuine Buddhists, the Burmans
are of course slaves to the strangest superstitions, and like the
Chinese, Malays, and other south-eastern Mongolic races, they are
one and all reckless gamblers. But, on the other hand, they **are**
generally of sober and frugal habits, their innate kindness, **good**
humour, and consideration for the feelings of others making **them**
general favourites with all who have any dealings with them.

Education, at least to **the** extent of reading the Buddhist texts and
writing their own language, **is** widespread amongst the men, most of whom
are brought up in the schools attached to the temples. The women also,
thanks to their social freedom, betray an unusual degree of intelligence
and aptitude for business. Hence it is probable that, once for all relieved
from the cruelty and exactions of the capricious and autocratic sovereigns
of the native Alompra dynasty, the Burmese nation will readily accept
European culture, and soon take a prominent part in the diffusion of
western ideas amongst the semi-civilised peoples of the Indo-Chinese

Peninsula. Here they form a compact nationality, which has long been
dominant throughout all the riverain parts of the upper and middle
Irawaddy basin, and which has gradually crowded out or absorbed the
Talaing (Mon) race, now nearly extinct as a separate ethnical element in
Pegu and the Irawaddy-Sittang delta.

The Talaings, who at one time held almost exclusive possession
of **this** region, from Prome to Maulmein, can no longer be distinguished physically from **their** Burmese neighbours. But their
language is totally different **from all the surrounding** idioms, and
affinities have been sought for it on **the one hand in** the Kolarian
of Central India, on the other, in the Annamese of Tonkin. The
natives recognise there divisions: the *Mon Tine* of Pegu, the *Mon Di*
of Rangûn and Tavoy, **and the** *Mon Myat Lawa* of Myawadi. The
term *Talaing* is the **same as** *Telinga* (*Telugu*), pointing at the
Indian origin, not of **the race,** but of its former rulers.

The Burmese national name, always written *Myamma*, and formerly
pronounced *Byamma* or *Bamma*, for *Brahma*, there being no letter *r* in
the language, is associated by the natives with the "nine Brahmas,"
from whom they claim descent. But it is obviously derived from a root
myo for *miro*, meaning "people," "men," a term by which some of the
primitive members of the race are still known in the Arakanese highlands.

THE SIAMESE, SHANS, AND LAOS.

Under a general uniformity **of** type the Siamese present **in**
their outward appearance and mental characteristics some marked
differences from their western neighbours. They are on the whole
a less vigorous race both physically and morally, of shorter stature,
and less robust frames, less independent and more subservient to
despotic rule. In Siam slavery, little practised in Burma, is a wide-
spread national institution, and the people, although in some respects
more cultured and refined, are at the same time more effeminate.
These differences may perhaps to some extent be accounted for by
the different origin of the two races, the Burmese coming directly
from the lofty Tibetan tableland, the Siamese from the low-lying
plains of Eastern China. Recent ethnological research has revealed
the fact that the Chinese people are not the primitive inhabitants **of**
the Yangtse-Kiang basin, which on their comparatively recent arrival
from the north they found already settled by a semi-civilised
agricultural race that has been identified with the modern Shans.

This term *Shan* is probably the same as *Siam*, which comes to
us through the intermediate Portuguese form *Sião*. But **in** any
case there can be no doubt that the Siamese are a southern branch
of **the great** Shan nation, the transition between the **two** being

effected by the *Lao* people of the Upper Menam and Middle Mekhong basins. All affect the general designation of Thai (Tai), that is, " Free," " Noble," and their long contact with the present inhabitants of the " Middle Kingdom " is shown by the constituent elements of the Chinese language, of which fifty per cent. are of Shan origin. The cradle of the Shan race has even been traced by Terrien de Lacouperie, with some show of probability, to the Kiulung highlands north of Sechuen and south of Shensi in west and north-west China.

But at present the chief home of the Shans proper are the borderlands between Yunnan, Burmah, and Siam. East of the Meping (Upper Menam), and generally in North and East Siam, they are grouped as Laos in two great divisions—the Lau-Phun-Ham, or " White-Paunch Lao," who do not practise tattooing, and the Lau-Phun-Dam, or " Black-Paunch Lao," who, like their Burmese neighbours, cover the body with wonderfully intricate tattoo designs, thus giving it a dark or black appearance. They are an historical people, who were formerly constituted in an ancient and powerful kingdom, whose capital, Vinh-Khianh (Vien-shan), was taken and destroyed by the Siamese about the year 1828. The western and northern Shans have also forfeited their independence to China, Siam, or Burmah, although the Shan country between North Siam and Yunnan (20° to 23° N. lat.) is practically autonomous. They are a semi-civilised people, engaged chiefly in trade and agriculture, with a knowledge of letters, and Buddhists, like all the settled populations of Indo-China. They have domesticated the elephant and buffalo, are peaceful and industrious, and skilled in the production of lacquered wares, and of silk and cotton fabrics for local use. Trading relations have long been established with China, Siam, Burmah, and Camboja, with which countries their ivory, gold dust, tin, gums, lac, benzoin, raw silk, skins, and sapan-wood are bartered for cotton cloth, chintzes, silk, opium, hardware, and porcelain. At present much of this trade is carried on by itinerant Shan and Burmese hawkers, who are met everywhere between the Irawaddy and Mekhong rivers, organised in small caravans, and well armed, like the Povindahs of Afghanistan.

Although nominal Buddhists, most of the Shans and Laos, and even many of the Siamese, are in reality still nature-worshippers, who make offerings of sticks and stones to the local genii, and guard their homes against evil spirits by means of brooms, cotton threads, bunches of herbage, or other curious devices. Some are quite as savage as the wild tribes, and although acquainted with the use of fire-arms, still use the national crossbow, a formidable weapon, which will kill a buffalo with a simple bamboo arrow at considerable distances. In some districts the confusion of types and usages is so great that the true wild tribes can be distinguished from the Shans and Laos only by the large bone, ivory, or wooden ornaments worn in the lobe of the ear, as amongst so many of the Oceanic, African, and American races. In European accounts of these wild tribes the confusion is increased by the generic designations mistaken for tribal names applied to them by their civilised neighbours. Such are *Moi* in Cochin-China, *Muong* in Tonkin, *Pnom* (*Penong*) in Cambaja, *Khâ* in the Lao districts, *Truo* in Lower Cochin-China, *Lolo* on the Yunnan frontier, all of which terms mean little more than savage, wild, or hill tribe in general,

and have no ethnical value whatsoever. The wild tribes are exposed to
the constant attack especially of the Lao princes, who organise regular
expeditions against them in order to procure slaves for the Siamese and
Cambojan markets. But apart from the passions fostered by this infamous
traffic, the Laos, like all the Thai-Shan peoples, are an inoffensive, unwar-
like, and peace-loving race, fond of music, and living chiefly on a diet of
rice, vegetables, fruits, fish, and poultry.

THE ANNAMESE.

Under the general name of Annamese are now usually comprised
all the settled inhabitants both of Tonkin and Cochin-China. They
constitute essentially one homogeneous people, nowhere presenting
any marked differences in type, speech, usages, or religion. Of all
the civilised nations of Indo-China they are certainly the least
favoured, both physically and morally, presenting so many disagree-
able traits that few observers have anything good to say for them.
To the Chinese they have been for ages known by the designation of
Giao-shi (Kiao-shi), occurring so early as the year 2285 B.C., and
supposed to mean "Bifurcated," or "Crossed Toes," from the
abnormal space between the great toe and all the others, a peculiarity
by which they are still distinguished. The Annamese are described
as the ugliest and most ungainly race in the peninsula, with a coarse
dirty yellow skin, broad head, flat, angular features, small snub nose,
thick lips, small pig eyes, and bow legs. The moral picture is
scarcely more flattering, and the Abbé Gagelin, who lived for years
in their midst, tells us that they are at once insolent and dishonest,
and dead to all the fine feelings of human nature. There is so little
affection amongst them that the nearest kindred never think of
embracing even after an absence of years. The missionaries are not
allowed to fondle the little children, nor is the slightest gesture
tolerated in the pulpit. M. Mouhot, who is nevertheless inclined to
speak well of them, confesses that "they are headstrong, revenge-
ful, deceitful, thieves and liars. Their dirty habits surpass anything
I have ever seen, and their food is abominably nasty, rotten fish and
dogs being their favourite diet." Hence on recently noticing the
absence of the Annamese element amongst the highland populations
towards the Chinese frontier, Mr. J. G. Scott not unnaturally ex-
claims: "This is satisfactory from one point of view. The fewer
Annamese there are the less taint there is on the human race."

Nevertheless, the same observer tells us that at least in one respect the
Tonkinese (Northern Annamese) are almost without rivals. They are
surprisingly skilful in the construction of embankments, and the dykes
built to guard the flat delta of the Song-koi against floods in the rainy
season are most admirably constructed—as a national work, far more astonish-

ing in the patient labour they imply than the Wall of China, or even than the Grand Canal itself. Other redeeming features are their love of home and of their native land. The children also, who are intelligent and fond of instruction, flock eagerly to the new French schools opened in Cochin-China, where most of the rising generation have learnt to read and write in the Roman character. Their own writing system is based on the Chinese, and, like their northern neighbours, they are merely nominal Buddhists or Confucians, the lettered classes concealing a scoffing spirit beneath fine moral maxims, the masses still worshipping the natural forces, ancestry, and the genii of the circumambient spaces. Amongst them the Roman Catholic missionaries had great success, especially during the 17th century, and notwithstanding subsequent persecutions and the recent wholesale massacres, there are still considerably over 100,000 Christians in the country.

Although polygamy is legally permitted, it **is** little practised except amongst the Mandarins and wealthy classes. On the other hand, divorce is so common that an interchange of wives may be almost regarded as a national institution. The Annamese are a short-lived people, a generally unhealthy climate, poor diet, and indolent habits, combined of late years with opium-smoking, causing them to age rapidly. Men fifty years old are already in a decrepit state, and few sexagenarians are met in the **country.**

THE CAMBOJANS.

The original stock of the Cambojan race are probably the rude Kuy people of western Camboja (province of Compong Soay) and south-eastern Siam (Mulu-Prey and Tonlé Repan), **to** whom the civilised Cambojans still give the title of Khmér-dom, that is, "Primitive Cambojans." The national name *Khmér*, in Siamese Kammén, has been identified with the Pali *Camboja*, an older form of which is *Kampushea*, according to M. Aymonier, the original name of the country. It is explained to mean the land of the Kam people, and in any case has nothing to do with the Kamboja of Sanskrit geography, with which it has been wrongly connected.

The Kuy aborigines are distinguished from the surrounding Mongolic peoples, both by their speech, which is untoned, and by their physical type, which may be described as almost Caucasic. The Cambojans proper also speak an untoned, polysyllabic language, which shows certain affinities with the neighbouring Cham, and with the more remote Oceanic group. But in their physical appearance the modern Cambojans have become through intermixture largely assimilated to the Siamese. They are descended of illustrious ancestors, who at one time ruled over a great part of Indo-China, and erected on the shores of Lake Tonlé-sap stupendous Brahmanic and Buddhist temples and other monuments, such as those of Angkor-Vat and Indapathaburi, rivalling in size and magnificence those of Java itself. But **the** modern Cambojans are a feeble,

r

decrepit race, unmindful of their past greatness, without national
aspirations for the future, satisfied to accept a present French
protectorate as the only escape from further encroachments on the
part of their Siamese and Annamese neighbours. Although more
honest, they are scarcely less indolent than the Annamese, whom
they also resemble in their unfriendly attitude towards strangers,
and in some other unamiable traits. On the other hand, they cultivate
the arts of music and poetry, accompanying their somewhat monotonous songs and duets on simple primitive instruments.

The ancient Cambojan culture, introduced and developed under Hindu
influences, seems never to have penetrated far below the surface. It failed
to eradicate many of the older usages, such as the practice of building the
houses on piles, which still largely prevails. In the different burial rites
are perpetuated the traditions and religious ideas of the several primitive
peoples merged in one nationality during the period of Cambojan prosperity. Some, especially of the poorer classes, burn their dead either
immediately, or three days after death ; others first bury and then disinter
the body, burning the bones years afterwards ; others again preserve it for
months and even years in their dwellings, injecting quicksilver, and
allowing the gases to escape through a tube which passes from the coffin to
the roof of the house. Polygamy, although legalised, is mainly confined
to the wealthy classes, and the women enjoy on the whole a considerable
share of respect and independence. They are even described as haughty,
jealous, and vindictive. Instead of ear-rings they often wear wooden, bone,
or ivory plugs in the lobe, which thus becomes distended to a monstrous
size. This custom, very general amongst many other primitive peoples in
various parts of the world, has persisted from pre-historic times in spite of
the foreign influences, under which were developed the Cambojan culture
and former political ascendancy in the lower Mekhong basin.

CHAPTER IV.

POLITICAL DIVISIONS—BURMAH, SIAM, ANNAM, CAMBOJA.

RECENT events have considerably simplified the political relations
in the Indo-Chinese Peninsula, which has now been brought under
the exclusive sway of one native and two foreign powers. The
three great political divisions thus constituted correspond very
accurately with the main physical divisions of the country. Thus
the British power, supreme in the west, comprises, besides the
coastlands on the Bay of Bengal, the Irawaddy and Salwin basins.
The French in the extreme east hold in the same way the Mekhong
and Song-Koi valleys ; while the central region, drained by the

Menam river, constitutes the native State of Siam. Certainly the boundaries between these political divisions are in many places ill-defined, or not at all laid down with any claim to accuracy. Nor do the great river valleys, always excepting the Menam, lie wholly within the respective territories, their furthest sources and head streams being found beyond the conventional frontiers, either on the Tibetan or the Yunnan plateau. But on the whole the tendency to bring the political and physical relations into complete harmony, has in recent times manifested itself as conspicuously in Indo-China as in most other quarters of the globe.

Amongst these political systems the peninsula is distributed in somewhat equal proportions—Burmah, in its widest extent, comprising nearly 300,000, Siam probably 290,000, and the French domain nearly 230,000 square miles. But if the estimated statistics can be trusted, the bulk of the population would appear to be concentrated in the eastern division, where the Song-Koi Valley (Tonkin) is said to contain about 12,000,000 souls, which would be more than half of all the rest of the peninsula, if the usual estimate of 6,000,000 for Siam could be accepted. Little reliance, however, can be placed on any statistical data, except from those English and French districts, where regular census returns are taken.

1. BRITISH DIVISION : BURMAH.

At the beginning of the present century the native State of Burmah, or Ava, as it was then called, comprised the whole of Indo-China lying between Siam and the Bay of Bengal east and west. Since then all the coastlands, including the three separate provinces of Arakan, Tenasserim, and Pegu, were successively ceded to England after the disastrous wars of the years 1825 and 1852, the independent territory being thus reduced to the position of an inland State. Lastly, towards the close of 1885, all that remained of the native kingdom was incorporated in the Anglo-Indian Empire. King Thebaw, last of the native Alompra dynasty, was deposed, and soon after the sovereign rights over his possessions of "Upper Burmah" were officially declared to be vested in the Kaisar-i-Hind.

Thus it happens that Burmah proper, with all its outlying dependencies, is once more united under one sovereign power, this political division of Indo-China again comprising exactly the same limits as it did before the war with England in the year 1825. The term "British Burmah," as distinguished from "Independent" or "Upper Burmah," has ceased to have any value, and on the latest maps the red line marking off British territory runs from Assam along the western and south-western frontiers of Yunnan eastwards nearly to Tonkin, and thence along the northern and western frontiers of Siam southwards to the Malay Peninsula. The region enclosed between this vast semi-circle and the Indian Ocean comprises three distinct physical divisions—Burmah proper, embracing nearly the whole of the Irawaddy and the greater part of the Salwin basin; Arakan, between Burmah proper and the Bay of Bengal; Tenasserim, between Siam and the same waters.

BURMAH PROPER.

Physical Features.—In this division must now be included both UPPER BURMAH, that is, the recently annexed native State, and LOWER BURMAH, that is, the district of PEGU, hitherto comprised with ARAKAN and TENASSERIM in the province of British Burmah. Between Upper and Lower Burmah there are no natural frontiers, and since the assimilation of the Talaing inhabitants of Pegu to the Burmese in speech and physical appearance, both form in all respects a perfectly homogeneous region, copiously watered by the great rivers Irawaddy, Sittang, and Salwin, and traversed by the parallel Arakan Yoma, Pegu Yoma, and Shan Yoma ranges.

Towards the north and north-west it is limited by the rugged and still little known Patkai highlands, separating it in this direction from the Brahmaputra basin. Nominally Burmah is here conterminous with the British province of Assam. But with the exception of the somewhat settled territory of Manipur, the intervening hills are mainly occupied by the Nagas, Lushais, Singpos, Khamtis (Kamptis), and other unreduced wild tribes or semi-civilised peoples not yet incorporated in the Anglo-Indian empire. The Patkai range was recently crossed by Col. Woodthorpe, who reached the settlements of the friendly Bor Khamtis on the western branch of the Irawaddy, after traversing some districts exposed to the constant raids of the unruly Singpo tribe. In the territory of the peaceful Kunnungs silver mines occur, which supply all the surrounding peoples with coin and ornaments. But just as the Khamti traders suffer from the attacks of the Singpos, the Kunnung communities are often plundered by the neighbouring Singlengs, who carry off the captives and sell them as slaves to the Tibetans. Near the village of Langdao the Irawaddy, here crossed by Col. Woodthorpe, was found to be only eighty-five yards broad and not very deep.

The Singpo and Khamti highlands merge eastwards in the Kakhyen (Kachyen) hills, which form the frontier towards Yunnan (south-west China), and which are held by the Kachyen wild tribes, over whom both China and Burmah have always claimed political supremacy. Since the annexation of Upper Burmah the rival claims of the two powers have led to diplomatic negotiations, by which it is hoped that the boundaries between the two empires may soon be clearly determined. Towards the west Burmah proper is bounded by the district of Arakan, and on the south it develops an extensive coast-line washed by the Bay of Bengal from Cape Negrais to Maulmain. Eastward it is supposed to be everywhere conterminous with Siam; but, except in Pegu, the frontiers are not very clearly laid down, and appear to have frequently shifted with the uncertain allegiance of the intermediate Shan States.

Within its conventional limits Burmah proper is comprised between ten degrees of latitude (26°—16° N.), stretching from the Patkai range for about 700 miles southwards to the Gulf of Martaban, with a breadth of over 400 miles at its widest part, and a total area of some 220,000 square miles. Of this space 190,000 square miles belong to the late kingdom of Burmah (Upper Burmah), and 30,000 to the British district of Pegu (Lower Burmah). The upper or northern section forms a hilly plateau of moderate elevation traversed in its entire length by the Irawaddy, and intersected by the lateral valleys of the Kyen-dwen (Chin-dwin) and Tapeng, the former flowing from the Manipúri hills in the north-west and joining the right bank of the main stream below Mandalay, the latter descending from the Kachyen highlands to the left bank above Bhamô.

The Pegu or southern section consists mainly of the Irawaddy-Sittang delta, a vast low-lying alluvial plain intersected by the innumerable branches, channels, and backwaters of the Irawaddy and Sittang, and during the rainy season exposed to frequent and widespread inundations. Since the British occupation extensive works have been undertaken to protect the plains from these periodical floods, which attain their greatest height towards the end of July and in August, when the discharge is sixteen or seventeen times greater than at low water in February and March. The head of the delta above Henzada, 150 miles from the coast, is now protected by a semi-circle of embankments, which skirt the left bank of the Nawun (river of Bassein), and the right bank of the eastern or main branch, which retains the name of the Irawaddy. Even above the delta a dyke 60 miles long follows the right bank of the main stream, intercepting the torrents from the Arakan-Yoma, and deflecting them to the Nawun branch. But these works have unfortunately had the effect of raising the bed of the river, and thus increasing the extent of the inundations, which in 1877 converted into a temporary lake a tract of some 500 square miles on the east side of the Nawun.

Arts and Industries.—Burmese culture, as represented by the local arts and industries, has always been far inferior to that of India and China. The artistic sense has found its chief expression in the practice of tattooing, which has been developed to an extraordinary extent, displaying considerable taste in the combinations of colour, and in the execution of intricate designs covering a large part of the body. But little progress has been made in the useful arts, which are restricted chiefly to cotton and silk weaving, pottery, metal work, and jewellery. The cotton fabrics, woven by the women on looms of a primitive type, are much inferior to those of India; and the silks, the raw material for which comes from China and the Lao country, are of a coarse texture, although certainly strong and durable. Everybody except the lowest classes wears silk, the finer qualities of which are imported from China. The earthenware often combines elegant forms with good quality, and the workers in metal produce rude cutlery, arms, and various implements, **and domestic** utensils in iron, tin, and copper. Gold and silver ornaments are much worn in all the large towns, but are more remarkable for richness and solidity than for delicacy of design and execution. A large government gun foundry, fitted with all modern appliances, was founded some years ago in the neighbourhood of Mandalay, but does not appear to have flourished under its French managers. In Pegu cutch, used for dyeing purposes, is manufactured for export, and here there are numerous steam mills for sawing timber and cleansing rice also for export. Boat-building employs numerous hands along the river banks, and in some districts salt and *gnapi*, a preparation of fish, are produced in considerable quantities. A good deal of paper is manufactured from the fibre of young bamboos, and the towns of Sillay and Nyung-u below Pagan are important centres of the lacquer industry. The Burmese lacquer-ware in some respects rivals that of Japan, and with improved methods of production would command a ready sale in the European markets.

Trade.—Hitherto most of the inland trade has been carried on with China chiefly through Bhamô on the Upper Irawaddy. Chinese caravans, in which the camel is replaced by the horse, mule, and ox, convey large quantities of raw silk and fine silken stuffs from Yunnan to this emporium, taking in exchange Burmese cottons, besides some Indian and British wares. The nearest Chinese station lies some

five or six days' march beyond Bhamô ; but hitherto all attempts to
establish a regular trade between India and south-west China by this
route have ended in failure. More success may attend these efforts
as soon as Upper Burmah is pacified, and the intervening Kachyen
border tribes reduced. But Col. Woodthorpe's expedition to the
Khamti country seems to show that the best and most direct trade
routes between the two empires will still be found to lie further
north, that is, from East Assam over the Patkai range directly to
the vast and populous province of Se-Chuen.

The trade of Lower Burmah, which has its chief outlets in the
ports of Bassein, Rangûn, and Maulmain, has acquired great expansion in recent years. Through these ports large quantities of British
and Indian wares are introduced into the Irawaddy basin, and thence
widely distributed throughout the peninsula. The chief articles
taken in exchange are rice and timber (ironwood, teak, and other
valuable woods). But to these staples of the export trade will soon
probably be added the petroleum, rubies, jade, cottons, and other
produce of Upper Burmah.

Geographical and Political Divisions.—For the geographical
and administrative divisions of the late kingdom of Burmah our
chief source of information is still Captain (now Colonel) Henry
Yule's account of the British Mission of 1855 to the court of Ava.
There are also extant two historical documents of great interest—an
inscription preserved in a temple near Ava, and another inscribed
on the great bell at Rangûn, the former giving a complete list of the
nine royal provinces with their several districts or territories as in
1650, the latter a list of the sixteen provinces with all their sub-
divisions in 1776 ; that is, after the maritime districts of Tavoy and
Tenasserim had been added to the empire by Alompra. Col. Yule,
who reproduces the Ava document, enumerates as under the more
important territorial divisions on the right or west side of the
Irawaddy basin :—

Hu-Khong, a rich valley about the upper course of the Kyen-
dwen at the southern foot of the hills towards the Assam frontier.
Here are some amber mines ; natives chiefly Kachyens, a branch of
the Singpo family.

Mogung, with a river and ancient city of like name, in the
extreme north-west, beyond Bhamô, between 25°—26° N. lat. This
is the *Mongmaorong* of the Chinese, peopled by the Kubo (Shan) tribe.

Muchobo, Alun Myo, and *Dabāien*, districts between the Irawaddy and the lower Kyen-dwen, due west and south-west of
Mandalay.

Ka'é Myo, town and district in the Kyen-dwen valley north-west of the foregoing.

Pakhan, on the Irawaddy, just below the Kyen-dwen confluence.

Yo or *Yau*, a large canton between the Irawaddy and the Arakan-Yotna hills, 21° to 22° N. lat., inhabited by the Yaus, of Burmese stock and speech, itinerant dealers, met everywhere in North Burmah.

Tsalen, south of the Yau territory.

Malûn and *Taindah*, south of Tsalén as far as the Pegu frontier.

On the opposite or left side of the Irawaddy basin the chief districts are :

Bhamô, with town of like name, a hilly country towards the Yunnan frontier, about the lower course of the Tápeng, and in-habited by Kachyen wild tribes, Burmese and Chinese.

Myadûng, Tagung, Tsampenago, Tsengu, Madey, districts following from north to south along the left bank of the Irawaddy between Bhamô and the capital.

Ava, Tarur, Myo, Pagan, Tsilé, Magwé, Myingun, districts following along the left bank from the capital to Pegu.

Yemesen, Nyenghyan, Tungdwen, west of the Sittang river as far as the Pegu frontier.

East of Burmah proper there are altogether about fifteen petty Lao and Shan States which have hitherto been subject to the crown of Ava, and which must consequently now be regarded as forming part of British Burmah. Nearly all lie beyond the Irawaddy basin, being watered by the Upper Sittang and the Salwin, and stretching east-wards along the north frontier of Siam to the Mekhong river.

West of the Salwin are *Mobye, Mokme, Moné, Nyang-yoré, Legya, Thein-ni, Thibo, Thung-zé*, and *Momeit*, administered from the town of Moné, former residence of the Burmese governor. They are collectively comprised under the name of KAMBOZA (KAMBOZA-TAING), a term of Hindoo origin not to be confused with the Cam-boja of the Lower Mekhong basin.

East of the Salwin are the six States of *Maing-leng-ghyi, Muang-ting, Kaing-ma, Liang-hung, Kiang-tung*, and *Kiang-khen*, which appear to have been comprised amongst the twenty-six royalties said to have been formerly subject to the kingdom of Pegu. Some have since transferred their allegiance to Siam, while, according to Carl Bock and A. R. Colquhoun, others have constituted themselves autonomous States independent alike of Burmah, Siam, and China. On the map of Indo-China accompanying Colquhoun's work, ' Amongst the Shans ' (1885), the late kingdom of Burmah is reduced

to the Irawaddy basin, the whole region stretching thence between
Siam and Yunnan eastwards to about 103° E. longitude being
indicated as the "Independent Shan Country." But this appears to
be going back to the thirteenth century, when the extensive Shan
empire embraced all the Kamboza States on the plateau between the
Irawaddy and Salwin rivers, besides many other parts of Indo-China.
And although the statement (p. 321) that "the Burmese Shan States,
which are now independent, contain about 80,000 square miles,"
may be an exaggeration, it seems certain that Kiang-hung and one
or two other Shan States have been independent both of China and
Burmah at least since the Mohammedan rebellion in Yunnan.

To these must be added the semi-independent territory of the
Karen-ni, or "Red Karens," which forms an enclave between the
Sittang and Salwin on the north-east frontier of Pegu. Like their
kindred in Tenasserim, many of the Karens have accepted the teach-
ings of the Christian missionaries, and will probably prefer the
British administration to the capricious government of their Burmese
rulers. But many of the wilder tribes between Lower Burmah and
Siam still lead nomad lives, and are described as "a frequent source
of trouble, committing highway robberies on British as well as
Siamese subjects" (C. Bock). Hence the Karen-ni territory, some
50 miles by 30 broad, has hitherto obstructed the regular trade
between Pegu and the Shan country west of the Salwin. Matters
were made worse by a treaty, in which both the Burmese and the
English agreed not to annex the district, the reduction of which
will certainly remove a great obstacle to the development of commercial intercourse between Burmah and the neighbouring Shan
States.

Topography.—Both in Upper and Lower Burmah nearly all
the large towns lie in the Irawaddy basin, and generally on the left
bank of that river. In the extreme north the only place of any
importance is *Bhamô*, just below the Tapeng confluence, a group of
600 or 700 houses protected by a stout bamboo palisade from the
raids of the surrounding Kachyen hillmen. One quarter is occupied
exclusively by Chinese artisans, and nearly all the overland trade
with Yunnan is in the hands of the local Chinese dealers.

Bhamô is distant some fifty miles to the north of the point, about
22° N. lat., where the Irawaddy bends suddenly westwards, and
where is situated the cluster of royal towns, *Sagain*, *Ava*, *Amára-
pûra*, and *Mandalay*, which have been the successive capitals of the
empire during the last 600 years. Nothing is so puzzling in the
history and geography of Burmah as this shifting of the imperial

residence, a practice which appears to have prevailed ever since the introduction of Buddhism some 400 years before the Christian era. During the early period the centre of authority lay far to the south, gradually moving from Prome through Pagan and Panya northwards to Sagain on the left bank, and thence in 1364 to Ava, at the confluence of the Myitngai at the opposite side of the Irawaddy.

Ava, which often gives its name to the whole country, remained the metropolis for nearly five centuries (1364—1837), except from 1783—1819, when it was replaced in the royal favour by Amárapúra, which stood five miles further north on the same side of the river. In 1837 the Court was again removed to Amárapúra, and in 1857 to the present capital, Mandalay, a little further north, but about two miles from the bank of the river. The two places are connected by a long line of houses, magazines, and dockyards, and both are laid out on the same plan, forming a regular square with brick walls, and in the centre a second quadrangular space containing the royal quarter. The mathematical centre of Mandalay is occupied by a seven-storied gilded tower marking the spot where stands the royal throne, representing the sacred Mount Meru of Hindu mythology. Malay lies so low that by the bursting of the embankment in August 1886, it was flooded to a depth of from 10 to 20 feet. It is reached by large steamers, while vessels drawing three feet ascend to the Tapeng confluence above Bhamô.

Below Mandalay there is no town worthy of the name until we reach *Myi Kyan*, opposite the Kyen-dwen confluence, in a rich rice-growing district, and at present the most flourishing trading place in Upper Burmah. The Kyen-dwen, by far the largest tributary of the Irawaddy, gives access from this place to the fertile plain of Manipur, a British vassal State near the Assam frontier, while its furthest head streams penetrate far into the Singpo and Khamti hills far to the north-west of Bhamô. A few miles below Myi-Kyan the left bank of the Irawaddy is strewn with the ruins of the historical city of *Pagan*, covering a space of over 16 square miles, and including nearly 1000 pagodas, many still in a good state of preservation, a statue over 160 feet long, and other objects of great interest. According to the local tradition the shrines originally numbered 9999, but some 6000 had to be destroyed in order to strengthen the defensive works when Pagan was besieged by the Chinese in 1284. This place, although founded about the year 850, is sometimes called *New Pagan*, to distinguish it from the still more ancient *Old Pagan*, 210 miles higher up the Irawaddy, which was already a royal capital in the second century of the new era.

Below Pagan follow *Yenan-gyong*, **centre** of the petroleum industry, *Menhla*, one of the few places of any importance on the right bank of the Irawaddy, and *Thayetmyo*, just within the former British frontier.

Facing Thayetmyo is the health-resort of *Allan-myo*, destined terminus of the Rangûn railway, which at present stops at *Prome* (*Promé*), one of the oldest cities in Indo-China. Prome, which lies in an extremely rich agricultural district, is said to have been founded about 480 B.C., and was already a royal residence in the third century before the present era. Although destroyed over 1700 years ago it soon rose again from its ashes, and is still the most flourishing place in Pegu north of the Irawaddy delta.

In the delta itself the chief centres of trade and population are *Bassein* on the western branch of the Irawaddy, 75 miles from **the** sea, probably the Besynga of Ptolemy, and *Rangûn* on the eastern branch, less than half that distance from the coast. Since its **occu-** pation by the British in 1852, Rangûn, present capital **of** British Burmah, has made rapid strides in material and social progress, and already ranks as one of the great centres of trade, population, and general culture in Southern Asia. At that time a mere collection of wretched bamboo hovels enclosed by a log stockade and fosse, it is now a stately city of over 200,000 inhabitants, well laid out with good streets, parks and gardens, fine warehouses, schools, hospitals, factories, and numerous public buildings. Confident in its future destinies, and already claiming the proud title of "Queen of the East," it forms the southern terminus of a railway which skirts the left side of the delta to Prome, and which will doubtless gradually creep up the Irawaddy basin to Myi-kian and Mandalay, if not to Bhamó, towards the north-west Chinese frontier ; it is frequented by large sea-going steamers, and is the centre of an ever-increasing import and export trade with all the surrounding lands and with England. The most noteworthy native monument is the famous pagoda of Shway Dagohn, yearly visited by thousands of pilgrims from the neighbouring Buddhist regions. It is a lofty structure, whose gilded spire rises to a height of nearly 400 feet above the ground.

Besides the Prome railway, another line is in course of con- struction, running from Rangûn north-eastwards to *Tung-nyu*, the chief town in the Sittang basin. The most important intermediate station is the once famous town of *Pegu* **on** the Pegu river, which communicates through separate branches westwards with the Ira- waddy, eastwards with the Sittang delta. But these channels being inaccessible to large vessels, most of the local trade has been diverted to Rangûn, and since its capture and destruction by Alompra in the last century, this ancient capital of the Talaings has sunk to the position of an obscure inland town. Its former importance, however, is attested by the fact that from it the whole of Lower Burmah takes the name of Pegu.

ARAKAN.

Physical Features.—The westernmost and second in importance of the three divisions of British Burmah, Arakan constitutes a clearly-defined geographical region confined east and west by the Arakan Yoma range and the sea, and stretching from the Chittagong division of Lower Bengal southwards to the watery plains of the Irawaddy delta. It thus forms a long narrow maritime zone, skirting the east side of the Bay of Bengal for 250 miles, from Cape Elephant, 21° 10′ N. lat., just above the Naf estuary, to Pagoda Point below Cape Negrais, 16° 2′ N. lat. In the north the Arakan Yoma range, separating it from Burmah, is distant from 80 to 90 miles from the coast, towards which it gradually approaches southwards, while diminishing from 6000 to 7000 feet in height in the same direction, until it merges at last in the Irawaddy delta, within 13 or 14 miles of Cape Negrais. Total area rather over 19,000 square miles.

Towards the centre the coast is fringed by a large number of islands, islets, and reefs, of which the largest are the volcanic Ramri and Cheduba. In the north also the seaboard is indented by several inlets forming the estuaries of the Myo, Naf, Koladyne, Lemru (Lemyu), and other streams, flowing in a southerly direction from the Arakan Yoma, and converging in a network of channels and low islands, through which the chief branch reaches the sea under the name of the Arakan river in 20° 5′ N. lat. Vessels of considerable size ascend the Koladyne, Myo, and Lemru for 30 or 40 miles, but the other coast streams are accessible only to the native craft.

Physically speaking Arakan constitutes little more than the western slope of the coast range, which forms the water-parting between the rivers flowing east to the Irawaddy basin, and west or south-west to the Bay of Bengal. This slope is much broken, especially in the north, by parallel terraces and deep river gorges densely clothed with magnificent forests of teak and other valuable timbers. Near the sea the soil is sandy, but about the Koladyne and Lemru estuaries the surface is occupied with extensive tracts of extremely fertile alluvial lands. In the interior also the argillaceous riverain valleys are found to be highly productive, wherever the land has been cleared of the dense growth of primeval jungle. But even since the British occupation in 1826 comparatively little land has been reclaimed and brought under regular cultivation.

Natural Resources.—Indigo, sugar-cane, and cotton are either indigenous or have been thoroughly acclimatised. More or less successful attempts have also been made to lay out tea plantations on the higher grounds, but rice will always probably form the chief agricultural crop in the well-watered lowlands. Melons, cucumbers, pine-apples, mangoes, and many other fruits thrive well, while tobacco promises to become a staple product in the northern districts, where it yields over 500 lbs. per acre in a soil so rich as to need **no** rotation of crops. Next to agricultural produce the chief resource of the country is its teak and other forest growths, which **have** developed a large and increasing timber trade, exported chiefly **from Akyab.**

Of mineral wealth there is very little, or very little has hitherto been discovered. Iron probably exists, and mention has been made of coal. Limestone also abounds, the prevailing geological formations being chalk and tertiary limestone and sandstone. Old plutonic rocks occur, but there is little trace of recent igneous action beyond the already-mentioned mud volcanoes of the Cheduba and Ramri islands. In this neighbourhood and in the Akyab district further north petroleum springs bubble up, and it is noteworthy that the petroleum region in the Irawaddy basin lies under the same latitude as Akyab. Here also there is distinct evidence of upheaval, and Round Island, lying between Cheduba and the mainland, is said to have been raised several yards during an earthquake in the middle of the last century.

Inhabitants.—The great bulk of the Arakanese natives belong undoubtedly to the same stock as their Burmese neighbours. They speak a Burmese dialect of a somewhat archaic type, and some of the tribes bear the common national name of *Mro*, that is, "*Men*," a word that has been identified with the Burmese *Myamma* (*Mramma*). In the Burmese chronicles the *Rakhaintha*, as the primitive inhabitants of the country are collectively called, receive the title of M'rammâkrih, "Great Mrammas," or Elder Burmese, and their traditions point to Rakhaing, that is, Arakan, as the cradle of their race.

The Rakhaintha are commonly divided into two groups—the *Tungtha*, or "Highlanders," and the *Khyungtha*, or "River People," that is, Lowlanders. The former, comprising the Mros,

Bangis, Pankhos, Kumis, and many other hill tribes, are still mostly in a wild state, living by the produce of the chase, and worshipping the powers of nature as manifested in all outward phenomena. In this respect, as well as in physical type, they appear to be closely allied to the Shendus, Nagas, Mishmis, and other wild tribes occupying the Chittagong and Assam highlands as far as the Tibetan frontier. The Khyungtha, comprising the Mugs (Mags), Khamis, and **other more** civilised communities settled on the middle and lower course of the rivers, constitute the agricultural element. Like the other cultured people of Burmah, they are chiefly Buddhists ; but **the** type has become considerably modified by intermixture with Hindu immigrants from Lower Bengal. Many of these immigrants belong to the low caste Mug tribe of east Bengal, whence the **term** Mug is now commonly but wrongly applied by the Bengalese to all the lowland or settled inhabitants of Arakan.

These settled populations, amongst whom are **a few** Mohammedans originally from Delhi and other parts of **the** Ganges **basin**, are engaged almost exclusively in agricultural pursuits. The local industries have either been arrested in their natural development or completely extinguished by the competition of Indian and European wares introduced from Chittagong, or through the port of Akyab. At this port are also shipped the rice and timber which form the chief articles of the export trade. Formerly a large transit trade **was** carried on with Burmah along a fine highway constructed by the King of Ava at the beginning of the present century over the intervening Arakan Yoma range. Along this route silks, cottons, and other European and Indian goods, besides such local produce as salt and betel nuts, were forwarded in exchange for ivory, silver, copper, precious stones, and other Burmese products. A railway constructed from Akyab along this route to Mandalay would tend to develop the vast resources of the Irawaddy basin almost more rapidly than a northern extension of the Rangûn-Prome line.

Administration.—Forming one of the three divisions of British Burmah as constituted before the recent occupation of Independent Burmah, Arakan is administered by a commissioner, who exercises the functions of civil and criminal judge, and controls all matters connected with the revenue, trade, and navigation. Under him are deputy commissioners in charge of the several districts. Under the native rule there were four districts, two on the mainland (Arakan proper in the north and Thandwai or Sandoway in the south), and two for **the** islands of Ramri and Cheduba. The number of districts is still the same, but the distribution is different, there being now three for the

mainland (North Arakan, Akyab, and Sandoway), and one for Ramri
and the adjacent islands.

Topography.—*Arakan*, the former capital, now known by the
name of *Wrobang*, or "Old Town," stood some fifty miles up a main
branch of the Koladyne in a fertile rice-growing district. But owing
to its unhealthy climate the seat of government has been transferred
to *Akyab* at the mouth of the Koladyne, which has become the chief
outlet for the trade of the country. Large quantities of rice are
shipped for Europe and India, and a considerable export trade is
also carried on in timber, especially ironwood, much used for railway
sleepers in India. It is a cheerful place, with several public buildings, and broad streets lined with fine trees. Since its occupation
by the English in 1827, Akyab, which the natives call *Tsettwei*, has
grown from an obscure fishing village to a large and flourishing
town of over 30,000 inhabitants. Further south are the small
trading ports of *Khyuk Hpyu* (Ramri), at the northern extremity
of Ramri Island, and *Sandoway* on the mainland below Cheduba,
both capitals of districts.

There are no other noteworthy centres of population in Arakan,
which, notwithstanding the progress made under British rule, still
continues mostly under dense primæval forest on the slopes of hills,
and in the lowlands under water during the south-west monsoon.
At this period almost the only dry road is the recently constructed
highway running from Sandoway across the hills to Prome in Pegu.
Being exposed to the full fury of the monsoon, with a rainfall never
under 120 and often exceeding 240 inches, Arakan, like the Assam
lowlands, must always depend on water as a chief means of communication throughout all the low-lying districts.

TENASSERIM.

Physical Features.—In its physical constitution, **Tenasserim**
presents in many respects a striking analogy with the northern
division of Arakan. It comprises the whole of the maritime region,
stretching from Pegu along the east side of the Bay of Bengal for
seven degrees of latitude (17°—10° N.) southwards to the Isthmus of
Kra. It thus stands in the same geographical relation to Siam on
its eastern border that Arakan does to Burmah; and as the Arakan
Yoma forms the divide between the Irawaddy basin and the Bay of
Bengal, the Tenasserim coast range separates the streams flowing
east to the Menam from those flowing west to the same marine
basin. **The** Tenasserim water-parting, however, is less regular and

also less elevated, seldom rising above 5000 feet, and approaching at some points close to the west, at others close to the east coast, that is, to the Gulf of Siam. The average distance from the Bay of Bengal varies from 30 to 40 miles, with a coast line of about 500 miles, and a total area of nearly 47,000 square miles.

The sea-board is even more diversified than that of Arakan, being broken by the estuaries of the Salwin, Tenasserim, Tavoy, and some other considerable streams, and thickly studded throughout its entire length by the innumerable islands, rocks, and reefs of the Moscos and Mergui Archipelagoes. These insular groups which skirt the coast for over 300 miles, appear to be the scattered fragments of partly submerged mountain ranges, running parallel with the inland range, and, like it, consisting of conglomerates, porphyries, and granites. The inland range itself must be regarded as a northern extension of the Malayan mountain system, and also abounds in tin, which is now worked by Chinese miners. Elsewhere stratified sandstones, interspersed with quartz veins, and containing crystals of great beauty, are a predominant geological feature, replaced in the north by extensive tracts of rich alluvial soil, and in the lower hills by laterite. Besides tin, other useful metals, such as lead, iron, copper, and antimony occur in the metalliferous districts of Mergui and Tavoy. Coal of an inferior quality has also been discovered in the lower Tenasserim river basin.

Although rising near the coast, the Tenasserim rivers acquire a considerable development by flowing, not directly to the Bay of Bengal, but in long valleys disposed mainly north and south parallel with the backbone of the country. Thus the Atteran flows north to the Salwin estuary, and the Tavoy winds south for about 120 miles to its mouth opposite Tavoy Island in the Mergui group, while the Tenasserim develops a total length of 300 miles during an erratic course, first north-west parallel with the Tavoy estuary, then south to the town of Tenasserim, and again north-west to its delta at Mergui, opposite King Island. The Tenasserim is navigable for about 100 miles, and the Tavoy estuary affords good anchorage for shipping.

Being exposed, like Arakan, to the full fury of the south-western monsoons, Tenasserim has an extremely moist climate, with a rainfall seldom less than 120 inches in the year, and often exceeding that amount. But notwithstanding this excessive moisture, the climate is not unhealthy on the hills, where the temperature ranges from 70° to 90° F.; even on the plains the glass seldom rises higher than 98° or 100° F.

The uplands are still covered with dense forest growths, chiefly of teak, sapan, ironwood, rattan, bamboo, and several species of gumniferous plants. Lower down the alluvial plains are well suited for the culture of cotton, indigo, tobacco, sugar-cane, rice, and all kinds of tropical fruits. But owing to the scanty population, scarcely fifteen per square mile, very little of the land has been reclaimed, and the primeval jungle still continues to afford a refuge to the elephant, tiger, rhinoceros, wild boar, and large numbers of deer.

The great bulk of the lowland population appears to be of Burmese and Talaing origin, speaking the Burmese language, and practising Buddhist and Jain rites. There is, however, a large intermixture of natives of India, probably not less than 40,000, who are in about equal proportions Mohammedans and Hindus, and who usually speak Bengali. The hills are still occupied by a few scattered aboriginal communities, mainly of Karen stock, conterminous on the east side with the Siamese, and reaching northwards to the kindred tribes in East Pegu and the Karen-ni country. Along the coast are met a primitive race of fishers known as Silongs (Selongs), who also occupy most of the Mergui archipelago, encamping during the monsoons on the islands, and at other times living in their boats or on the beach. They appear to be an outlying branch of the Malay race, in the same low state of culture as the Orang-laut or seafaring Malays of the pre-Mohammedan epoch.

Topography.—By far the most valuable part of Tenasserims, the northern division of Amherst, which borders on East Pegu, and comprises the fertile alluvial plain of the Lower Salwin and its delta. Here is situated the formerly important but now much reduced town of *Martaban*, which gives its name to the neighbouring gulf. Facing it on the Salwin estuary lies the present capital, *Maulmain*, a flourishing seaport, sheltered from the south-west monsoons by the adjacent island of Belu, or Belugyun (Bhilu-ghaiwon). The motley Burmese, Indian, Chinese, and European inhabitants of Maulmain are mostly engaged in trade, exporting rice, teak, cotton, and other local products in exchange for European and Indian wares. Some thirty miles down the coast lies the little health resort and watering-place of *Amherst*, which gives its name to the northern district of Tenasserim, and affords a refuge to. the rich traders of Maulmain during the oppressive summer heats.

The only other noteworthy centres of population are the small inland town of *Tenasserim*, on the lower Tenasserim, whence both the river and province take their name, and the small ports of *Mergui*, on the delta of the same river, and *Tavoy* at the head of

G

the Tavoy estuary. Both have steam navigation with **Maulmain,** Rangún, and Calcutta.

Administration.—From the time of its occupation by the British in 1828 till 1862 Tenasserim was governed by a separate Commissioner under the Supreme Government. But in 1862 the three divisions of British Burmah were united under a Chief Commissioner, dependent on the Government of India, but with full control over **all** local departments. Since then Tenasserim, like **Pegu** and Arakan, has been administered by a Commissioner and **Deputy** Commissioners, **subordinate to** the central authority at **Rangún.** For administrative **purposes it forms** six districts, with forty-one sub-divisions.

2.—NATIVE DIVISION. SIAM.

Encircled west, north, and east by **the British and French** divisions, the native territory of Siam **occupies the** very **heart of** Indo-China, with **a** southern seaboard sweeping round the **Gulf of** Siam from Malay-land to Camboja. The south-western portion, comprising the Isthmus of Kra, and the Siamese section of Malay-land, and usually spoken of as LOWER SIAM, has been described in *Part 1. Malay Peninsula,* of which region it forms a natural geographical division.

UPPER SIAM, or SIAM PROPER, comprises the whole of the Menam basin, and the section of the Mekhong valley lying between Upper Burmah and Camboja, besides the valleys of the smaller streams flowing to the left bank of the Lower Salwin. It forms a compact, irregular square mass stretching from Kiang-tsen on the Mekhong, across over eight degrees of latitude (20° 15′—12° N.), for 560 miles southwards to the Gulf of Siam, and for about the same distance west and east between the Salwin and Cochin-China, with a total area of somewhat less than 300,000 square miles. The estimates of population vary enormously from 7,000,000 to four or five times that number. Carl Bock, who recently travelled through the most densely populated part of the country, from Bangkok up the Menam valley to the Mekhong and Kiang-tsen, is inclined to believe that the 7,000,000 of Pallégoix and others represents only the male adult population, " women and children under age not being counted." Mr. Colquhoun also, a still more recent explorer, clearly shows that the country is far more densely peopled than is generally supposed, so that the estimate of 25,000,000 made by the Siamese Ambassador in London may be accepted as probably not far from the truth. The constituent elements are the Siamese proper in the southern provinces, the kindred Shans and Laos in the north-western and north-eastern districts, some Cambojans, half-assimilated to the Siamese type, towards the Cambojan frontier, and the Chinese, who are very numerous, especially in the capital, and about the lower Menam valley.

Physical Features.—Except in the central parts, watered by the middle and lower Menam, Siam is essentially an upland region, everywhere diversified by isolated hills, broken ridges, or long mountain ranges. A prominent feature of this irregular and little known orographic system is the transverse Dong Phriya-fei chain, which north of Siam proper runs east and west, intersecting the course of all the streams flowing southwards to the Menam. Here the Meping, or Upper Menam itself, traverses a deep rocky channel, which in a space of about 90 miles between Zimmé (Xieng-mai), and Raheng is interrupted by over thirty rapids, mostly, however, accessible to steamers at high water. The elevated land presents the general aspect of plateaux all disposed north-east and south-west between the river valleys, but nowhere exceeding 3500 feet in height, except close to the Yunnan frontier. From the lofty summit crowned by the city of Xieng-tung the eye sweeps over a boundless prospect of peaks and domes covering a great part of the Lao country. From this point branch off two ranges, one running at a height of about 5000 feet between the Menam valley and Burmah southwards to Tenasserim, the other towards the Battambang and Parsat highlands between Siam and Camboja.

In the southern and eastern Lao country, along the course of the Mekhong, the relief of the land has been largely caused by the upheaval of the Devonian system subsequently to the development of the thick triassic deposits resting on the limestone formations. Further modifications are due to the eruption of the porphyries cropping out here and there, to post-triassic denudation, to the erosive action of running waters, and to alluvial deposits.

The alluvial formation finds its widest development in the great central plain of Indo-China, which constitutes Siam proper, and which is abundantly watered by the Menam and its numerous affluents from the east and west. Rising in the Shan country, near the converging point of the Salwin and Mekhong basins, at a height of some 900 feet above sea-level, the Meping, or western branch, usually but wrongly regarded as the true upper course of the Menam, flows mainly in a southerly direction to Raheng, where it is joined by the Menam-vang, a large affluent from the north-east.

The united stream, which now first takes the name of Menam, trends from this point south-eastwards to Paknam-Po, about 16° N. lat., where its volume is more than doubled by the Menam-yai, or "Great Menam," which comes also from the north-east, and which is rightly regarded by the natives as the true upper course of the Menam. The Menam-yai, or eastern branch, is described by Colquhoun as much larger and better navigable than the Meping, or western branch ; and as its valley lies also in the same direction as that of the lower Menam, it seems to be in every way entitled to be considered the main head stream of the great Siamese artery. It rises in an almost unknown region, enclosed north and east by the great bend of the Mekhong, and after collecting the waters of the Pré, Nan-kot, and other unexplored affluents, pursues a winding course through a fertile and thickly peopled country to the confluence. Beyond this point the main stream continues its southerly course, meandering sluggishly through a more open region, which gradually assumes the aspect of a rich, alluvial, low-lying plain, forming one of the great rice-growing districts of Further India. But from the banks of the river itself little is seen of the cultivated paddy-fields, which are in many places entirely concealed by a tangled growth of palms, bamboos, and other tropical vegetation fringing both sides of the stream.

As they approach the Gulf of Siam the united waters of all the Menams develop an intricate system of channels and backwaters, all subject to wide-spread floodings during the periodical summer risings. To these annual inundations the fertility of the soil is mainly due. Even as far as the Lao States the water rises from eight to ten feet during the rainy season, and, as in the Nile valley, an insufficient rise would be followed in Siam by a corresponding failure of the rice crop.

Towards the Menam delta converge two large streams, the Meklong from the north-west, and the Bang-Pak-Kong from the Korat highlands in the north-east, both reaching the head of the gulf about 20 miles to the east and west of Bangkok respectively, and both connected with the Menam by artificial or natural canals. The alluvial character of this region, which in some places contains extensive permanent swamps or lagoons, often overgrown with tall grasses, and frequented by numerous herds of wild elephants, is clearly shown by the borings for a well sunk in Bangkok to a depth of 25 feet through marine beds abounding in sea-shells and crustacea. The sea, which evidently at one time penetrated far inland to the foot of the Korat hills, has been gradually encroached upon by the sedimentary matter washed down with the numerous streams converging in the Menam delta. As the movement still continues, the time is approaching when the narrow inlet at the head of the Gulf **of Siam** will be entirely filled in, and when it will be possible to pass overland from Mergui in Tenasserim directly eastwards to Shantabun in south-east Siam.

As seen from the summit of the isolated Mount Patavi some 50 miles north-east of the capital, the eastern section of Siam, draining towards the Mekhong basins, presents a striking contrast to the level or slightly undulating plains of the middle and lower Menam valley. The view from Patavi, which stands over against Prabat, the sacred mountain of the Siamese Buddhists, reveals a vast prospect of rugged highlands to the north and east, and towards the south-east merging in the Xong and Pursat coast ranges between Siam and Camboja. But although crossed at different points by MacLeod, Bastian, Mouhot, and a few other European explorers, this upland region between the Menam and Mekhong still continues to be one of the least known tracts in the peninsula. The forest-clad hills are mostly occupied **by** independent wild tribes exposed to the constant attacks of the more civilised **Lao** communities, who organise regular slave-hunting expeditions **to supply** the slave-markets of Bangkok and Camboja.

The seaboard, which, including Lower Siam, develops **a** vast semi-circle of about 1000 miles round the Gulf of Siam, is mostly of a monotonous character, destitute of any deep bays, inlets, or other prominent features, and broken only by the mouths of the Menam, Shantabun, and a few other streams. Along the coast the depth varies from 40 to 60 feet, increasing to over 350 in the centre of the gulf, with a current of three miles an hour, which sets from south to north during the southern, and in the opposite direction during the northern monsoon.

Climate.—As in the Malay Peninsula, these alternating monsoons determine the distribution of moisture and the general aspect of the climate. The dry north-easterly trade-winds, prevailing from October to May, are followed for the rest of the year by the rain-bearing south-western currents, with a mean annual rainfall of 60 to 70 inches. Owing to the invigorating character of the northerly breezes, the climate is on the whole fairly healthy, the malaria on the low-lying coast-lands being less virulent than in the Ganges delta and other parts of south-eastern Asia. Although in the lower Menam valley the mean temperature is over 80° F., the heat is seldom oppressive except in the spring months towards the end of the northern and beginning of the southern monsoon.

Products and Natural Resources.—Of forest growths the most valuable are teak, sappan, eaglewood, garcinium yielding the gamboge of commerce, two species of cardamom, gutta-percha, lac, and several other gummiferous varieties. Amongst cultivated plants the leading place is taken by **rice,** which supplies the great staple **of** food, and of which as many as forty varieties are found. Next in importance is cotton, which seems to be indigenous in the upland regions, and which after supplying the native looms is exported to China. Other economical plants are sugar of excellent quality, tobacco **widely** cultivated on the plains, maize, the cocoa-nut and areca palms, **black** pepper in the Shantabun district, the soybean and ground pea. Fish, a chief article of diet, abound in the rivers **and** gulf, and large quantities of ngapi, a favourite dish at every **table, are** prepared from fermented fish and shrimps. Much bee's-wax comes from Battambang.

The mineral resources are chiefly copper, tin, magnetic iron ores, and antimony in the Lao country north of the Korat mountains. Several streams are washed for gold, and rubies, sapphires, and other precious stones are sent down by the torrents from **the** Koh-Sabap, or "Mountain of Gems," east of Shantabun. The gold mines or auriferous sands of Tu'k Sho in Battambang **are** also **productive.** They are at present worked by two Chinese companies.

Industries—Trade.—In the useful arts of life the Siamese have scarcely advanced much beyond the somewhat low level **of** their Burmese neighbours. From the native cotton they weave fabrics of coarse and finer texture for the local requirements, and also prepare their own earthenware. Formerly both the Siamese and Laos displayed considerable skill in bronze casting for the Buddhist temples; but at present the smelters and workers in iron appear **to** be mainly the resident Chinese. The national taste and inventive faculty have been chiefly exercised in **the** design and structure of their sacred edifices and royal palaces. The finest monuments of past times are found in the ruined city of Ayuthia. But the group **of** temples visited by Crawfurd in 1821 still covered a square of 650 feet on each side. The temples within the enclosure, disposed round a large central building, contained altogether 1500 images of Buddha, some of which were of colossal size. The ground storey of these structures is usually of plain brick and mortar, all ornamentation being reserved for the elaborately carved upper portion **and** teak roof, richly gilt on both sides, or covered with a coat of bright vermillion. The etchings also, whether of brass, bronze, or brick, are usually gilt all over. Some of the large effigies of Buddha stand beneath lofty pyramidal spires attached to the temples, the most noteworthy of which **still** towers to **a** height of **some** 400 feet amid the rank tropical vegetation concealing the silent ruins of Ayuthia.

Occupying the centre of the peninsula, with a seaboard lying midway between the Bay of Bengal and the China Sea, Siam is admirably situated for the purposes of international trade, and whenever the Isthmus of Kra is pierced by a navigable canal Bangkok must become one of the great emporiums of the Asiatic world. A considerable traffic has for ages been maintained overland through the Menam valley and Zimmé with China, and by water with all the surrounding insular and peninsular lands. The chief staples of export are rice, pulse, cocoa-nut oil, resin, cardamoms, pepper, gamboge, sappan, dyewoods, teak, eaglewood, indigo, stick-lac, ox and buffalo hides and horns, ivory, salt fish, and salt. The salt is partly mineral and partly derived from extensive salines at the mouth of the Meklong. In exchange for these commodities Siam imports tea, raw and woven silks, paper, earthenware from China; opium and cotton fabrics from India; hardware, machinery, arms, cutlery, glass, woollen and cotton goods from Europe.

The trade of the fertile Shan and Lao districts is with Yunnan to the north and Cochin-China to the east, to which it sends elephants, precious stones, gold, musk, ivory, wax, stick-lac, bamboos, cotton, and the so-called "Pegu ponies," which are really bred in the Shan country. In exchange are imported salt, fish, oil, silken stuffs, fire-arms, and gunpowder. There is also some trade with Burmah to the west, which will probably be much developed with improved communications, and the introduction of law and order into that hitherto distracted region.

Communications.—At present the only means of inland communication are the great waterway of the Menam, the trade route leading thence north to China, and the forest tracks over the hills east to Camboja, west to Burmah. Recently, however, the whole of the Menam valley has been carefully surveyed by Mr. Colquhoun, who strongly advocates a railway from Bangkok through Rahine (Raheng), to Zimmé, to be ultimately continued along the historic "Golden Road" through Zimmé to Esmok, within the Chinese frontier, and with a branch from Rahine westwards to the Burmese railway system. The ground has been examined and plans drawn out for these works, which might be executed as far as Kiang-tsen on the Mekhong at an estimated cost of about £3,288,000. No engineering difficulties would be encountered as far as Rahine, and few above it, as the line would run mainly through a plain studded with villages, and with a fall of not more than 900 feet for 500 miles from the north frontier of Siam to the capital. "The natural richness of the country," writes Mr. Colquhoun, "is remarkable, and when the railway from Bangkok to Rahine is built it will certainly be carried on to Zimmé in a short time. Whoever has not visited this place can with difficulty estimate the importance which the trade has already won, and how great its future prospects are. The government of India and the commercial community of this country are now fully alive to the importance of the question; and with the cordial assistance of the King of Siam, a reconnaissance of North Siam and the Shan country will shortly be made, as the initial step of this policy."

Political Divisions.—Like the late kingdom of Burmah, Siam comprises regions partly administered directly by the central government, partly held by the looser tie of vassalage and real or nominal tribute. To the former category belongs Siam proper, including the greater part of the Menam valley, and the provinces during the present century wrested from the kingdom of Camboja; to the latter the Shan and Lao States occupying the northern section of the Menam basin between 17°—20° N. lat., and the region stretching thence eastwards to Annam.

Siam proper comprises forty-one separate provinces, each governed by a phya, or functionary of the highest rank, appointed by the central government. They are distributed as under:—

Northern provinces, five : *Saug-Kalok*, *Phitsanulok*, *Phixni*, *Raheng*, and *Kampheng Pet*.

Eastern provinces, ten : *Pechabûn*, *Bua-Xum*, *Saraburi*, *Pachin*, *Kabin*, *Naphaburi*, *Nakhon-Nayok*, *Susang-Sao* (*Petrui*), *Battabong*, and *Phantaini-Khon*.

Western provinces, seven : *Muang-Sing*, *Suphanaburi*, *Kan-Sharanburi* (*Pak-Phrek*), *Racaburi*, *Nakhon Xaisi*, *Sakhonburi* (*Tha-Shin*), *Samut-Songkram* (*Meklong*).

Central provinces, nine : *Pak-pret*, *Nontaburi*, *Paknm-mutual*, *Ayuthia* (*Krung-Kao*), *Ang-Thong*, *Muang-Phrom*, *Muang-In*, *Xainat*, *Nakhon Sawan*.

Southern provinces, ten : *Paklat*, *Paknam*, *Banpplasoi*, *Rajong*, *Shantabûn*, *Thung-jai*, *Pechaburi*, *Xumphon*, *Xaiya*, and *Xalang*.

The Cambojan provinces now under Siamese rule, and also administered by high Siamese functionaries, are as under :—

Battambang, west of the Great Lake, with extensive Buddhist remains at Basset and other places ; centre of the bees'-wax industry.

Korat, a large and rich district north from Battambang.

Angkor, on the north side of the Great Lake, where stood the ancient capital of Camboja, whose site is still marked by the stupendous ruins of Angkor-Vaht.

Shukan, north-west of Angkor, also containing some remarkable ruins, such as those of Bantey-Shumar and the great bridge of Stung-Streng.

Suren, a forest country north of Shukan.

Sanken,
Kukau,
Melu-prey,
Tuly-repu, } wooded and hilly districts to the east of Suren and Shukan ; still largely occupied by wild tribes, and very little known to Europeans

The Lao country is partly tributary, partly divided into provinces administered directly by Siam. According to Bock, there are at present six Lao States directly tributary to Siam: *Lakon, Lampân, Chiengmai, (Xiengmai* or *Zimmé), Muang-nan, Hluang Prabang,* and *Muang Prai.* These lie chiefly in the north between the Mekhong and Salwin, thus including all the head waters of the Menam.

Of the Lao provinces ruled by Siam, often through governors chosen from the ancient royal family, the chief are: *Ubon* and *Bassac,* whose rulers bear the title of king; *Kemmarat, Noup-kai, Xieng-kang, Ken-tao, Xieng-kong,* and *Xieng-hai.*

Administration.—All the Lao States are absolutely independent of each other, their rulers being autocratic and elected for life, although the office is not hereditary, but filled by the King of Siam on the recommendation of the people. Each of these States has two chiefs, like Siam itself, the first with the title of Chow Hluang, the second called Chow Operat. The tribute to Siam consists of presents, such as gold and silver trees, gold and silver betel-boxes, ruby-studded vases, and the like, paid triennially.

Although bearing the official designation of Muang-Thai, or "Land of the Free," Siam proper is essentially a land of abject servitude. Theoretically the whole population, from the highest official to the lowest subject, are slaves of the Crown, whose power is limited only by custom. A number of distinct classes are, however, recognised from the nobles and military down to the slaves properly so-called. The political power is reserved to the nobles, the highest authority being vested in two kings, one paramount, the second with the title of Wagna, subordinate and nominal head of the army. These offices are hereditary within the royal dynasty, but do not necessarily pass to the eldest son. The present king, whose full title is Phra Bat Somdeth Phra Paramindr Maha Chulalonkorn Phra Chula Chom Klao Chow Yu Hua, is the ninth son of his father and predecessor, King Mongkut, whom he succeeded on October 1, 1868, in his fifteenth year. He is a wise and enlightened sovereign, who has profited by the liberal education which his father was careful to give him. His earnest desire to elevate the social condition of his people was shown by the abolition of slavery in his immediate dominions, an act which began to take effect in 1872. The children of slaves are free, and other important social reforms are in progress.

Topography.—*Bangkok,* the capital of **Siam** since the year **1769,** lies on both banks of the Lower Menam, about 20 miles from the sea, in 13° 18′ N. lat., 100° 34′ E. long., the river being navigable to this point for vessels of 350 tons. The striking appearance of the city, with its picturesque buildings and large floating population, has earned for it the title of the "Venice of the East." The palace of the "First King," enclosed by lofty white walls over a mile in circumference, forms a group of detached structures, such as temples, public offices, seraglios, the stalls for the sacred elephant, an arsenal, a theatre, and accommodation for some thousands of soldiers, cavalry and infantry. The hall of audience lies in the middle of the chief court, and in one of the temples is the famous jasper statue of Buddha. The population, estimated at over 400,000, includes natives of Burmah and Camboja, Annamese, Malays, Laos, Indo-Portuguese, and Europeans, besides the predominant classes of Siamese and Chinese. Bangkok is the commercial centre of the whole kingdom, the chief articles of export being rice, sugar, pepper, forwarded in exchange for European manufactures. Of late years machinery has been extensively introduced, and steam-mills set up for various purposes. Gas is used in the royal palaces, and houses of many of the nobility. A considerable number of European firms carry on business in the city, which is the seat of a Roman Catholic bishop, and the headquarters of several Christian missions.

Ayuthia, the former capital, lies about 45 miles farther up the river. It was founded in 1350, a date marking the dawn of authentic Siamese history, and was destroyed by the Burmese in 1766. It is now partly laid out as an elephant park, round which are strewn the remains of temples and other monuments embowered in the foliage of a rank tropical vegetation, but still attesting its former splendour.

On the coast the chief seaport is *Shantabun*, some miles southeast of the capital, with a mixed Siamese, Chinese, and Annamese population, and a considerable export trade in pepper (25 piculs = 3350 lbs. yearly), precious stones from the neighbouring Xong district, some cardamoms, and ivory. The French are endeavouring to establish a line of steamers between Saigon and Bangkok, touching at Shantabun.

In the interior are several important places, many of which, from the reports of the latest explorers, appear to be more extensive and populous than had hitherto been supposed. Such are *Korat*, the largest town in Siamese Camboja, east from Mount Patavi, near the source of a western affluent of the Mekhong; *Xieng-Kang, Hluang-Prabang, Xieng-Kong*, and *Xieng-tuen*, on the Mekhong, the last-named being the northernmost town in Siamese territory; *Raheng* in the Menam valley, north from Bangkok, a probable future centre of railway communication between Siam, Burmah, and Yunnan; lastly *Zimmé*, or *Xieng-mai (Cheng-mai)*, the great capital of the Siamese Shans, on the Meping, or Upper Menam, 180 miles north of Raheng, and 500 miles from the capital. Zimmé, which lies on the right bank of the Meping, about 800 feet above sea-level, is by far the largest and most flourishing place in the interior, comprising an inner and outer town, each with separate fortifications, and a total population of probably over 100,000. Like Raheng, it is destined to become a great centre of railway traffic in the near future.

Historical Notes.—Siam appears to have no place in recorded history prior to A.D. 638, and the authentic annals date only from the foundation of Ayuthia, the old capital, in 1350. The first notice of the country by European writers occurs in an account of an overland expedition against Malacca in 1502. In 1612 an English ship ascended the Menam as far as Ayuthia, and eight years afterwards the Portuguese sent thither their first missionaries.

In 1683 Constantine Phalcon, a Cephalonian Greek, became prime minister, and introduced some knowledge of European culture. Mutual embassies were exchanged at this period between Siam and France, Louis XIV. sending in 1685 the celebrated M. de la Loubère, to whom we are indebted for a graphic description of the country. But the expulsion of the French in 1690 was followed by a long period of civil strife and disastrous foreign wars, during which Alompra, founder of the late Burmese dynasty, seized Martaban, Tavoy, and Mergui, and overran the whole Menam basin. The loss of the Tenasserim provinces in 1759 was, however, compensated early in the present century by the extension of Siamese rule over a large portion of Camboja and the north Malay States of Kedah, Patani, Ligor, and Kelantan.

About the middle of the eighteenth century a Chinese mandarin succeeded by bribes in obtaining a monopoly of nearly every article of commerce, the result being the almost total extinction of trade. To remedy this evil, which was severely felt, especially by Great Britain and France, Sir John Bowring, British plenipotentiary in China, was commissioned to the Siamese court to induce the king to grant free trade. The result of the embassy was a treaty of commerce concluded in 1855 between England and Siam abolishing all monopolies, opening up the trade of the country on liberal principles, guaranteeing the security of European traders, and establishing a British consulate in Bangkok. In 1856 a commercial treaty was also concluded with France, and since then more intimate relations have been established with these two rival western powers, which seem to foreshadow the ultimate partition or extinction of the ancient kingdom of Siam.

3. FRENCH DIVISION : ANNAM AND CAMBOJA.

General Survey.—The French, or eastern, presents several points of analogy with the British, or western, division of the peninsula. Both are overshadowed in the north by the lofty plateaux of southern

China, from which they project southwards, presenting a coast-line of from 1000 to 1200 miles, the one towards the Bay of Bengal, the other towards the China Sea. Both also are traversed by extensive coast ranges, breaking up each region into several distinct physical sections, and watered by two great rivers flowing southwards and terminating in the vast Irawaddy and Mekhong deltas respectively. This curious parallelism extends even to the ethnical and political relations, the western Burmese and Talaings finding their counterpart in the eastern Annamese and Cambojans, all of which elements have been only quite recently and almost simultaneously brought under the direct sovereignty of the two powers contending for absolute supremacy in the Indo-Chinese peninsula. But here the analogy ceases, or is rather replaced by a violent contrast, for while British Burmah can be regarded only as a natural and almost inevitable expansion of the adjacent Indian Empire, the French conquest seems like an aggressive movement, unmotived by any political urgency, commercial or colonial interests.

As in the west since the extension of British rule to the whole of the Irawaddy basin, the distinction between Upper and Lower Burmah has ceased to have any significance, so in the east the French occupation of the whole region has much simplified its somewhat intricate geographical and political nomenclature. By the forced retirement of the Annamese emperor, and the king of Camboja's voluntary surrender of his regal functions, the terms *Tonkin* and *Cochin-China* as opposed to *Annam*, *Lower* or *French Cochin-China* as distinct from *Camboja*, have ceased to possess much more than an historical value, and the whole region thus merged in one political system may now be conveniently treated as a single geographical area.

Position—Extent—Population.—Taken in this comprehensive sense, FRENCH INDO-CHINA, as it must now be called, describes an extremely graceful curve in the form of a letter S round the China Sea, the two extremities expanding into the great deltas of the Red River and Mekhong (see p. 54), while the connecting shaft is formed by the intermediate coast range. It thus comprises three separate geographical areas—the alluvial plains of Tonkin and Camboja (with Lower Cochin-China) in the north and south, and the coast region of Cochin-China proper in the centre. Bounded on the north by the Chinese provinces of Kwang-si and Yunnan, on the west by Siam, and elsewhere by the China Sea, it stretches across fifteen degrees of latitude (23° 30′—8° 30′) for about 1000 miles between the Chinese frontier and Cape Camboja. In breadth it varies enormously, from over 400 miles in the south (103°—109° E. long.) and 280 in the north (102°—108° E. long.) to little over 50 in the central parts, where the Siamese frontier advances to a short distance of the China

Sea. With a coast-line of over 1500 miles, sweeping round from
Cape Paklung on the Gulf of Tonkin to the Siamese frontier on the
Gulf of Siam, it is so contracted in the central parts of Cochin-China
that the total area cannot be estimated at much more than 220,000
square miles, with a population of probably less than 20,000,000.
The great bulk of the inhabitants, say five-sixths altogether, belong
to the Annamese stock, the rest being chiefly Cambojans and Malays
in the lower Mekhong basin, Chinese numerous in all the trading
places, and wild tribes, here collectively known as Moi and Muong,
thinly scattered over all the hilly districts of the interior.

Physical Features.—The carefully cultivated and abundantly
watered alluvial plains of *Tonkin*, studded with large towns and
populous villages, and everywhere intersected by natural or artificial
canals, present a striking resemblance to the more prosperous provinces of the Chinese empire about the lower courses of the Hoang-
ho and Yangtse-Kiang. Hence this region, mainly comprised in the
Song-koi and Thai-binh basins, has often been regarded as little more
than a southern extension of China proper, of which it has for ages
formed an outlying tributary or protected State, and to which it has
been indebted for its arts, letters, religion, and general culture.
Nevertheless, Tonkin is completely severed from the Middle Kingdom by an almost unknown highland region, which appears to merge
gradually in the Kwang-si and Yunnan tablelands, but which, in
accordance with the jealous policy of its northern neighbour, has long
constituted a sort of neutral zone between the two States. This is
the home of several semi-civilised Muong tribes, who are commissioned by the imperial government to guard the two chief frontier
passes of Bien-Kwang and Nam-Kwan, and otherwise prevent all
regular intercourse between the populations on either side. This
rugged tract, through which the head-streams of the Song-koi flow in
deep rocky beds down to the plains, is also the refuge of the unruly
elements of Southern China, and notably of the famous "Black
Flags" who figured so conspicuously during the recent military
operations of the French in Annam.

The irregular range of hills skirting the Tonkin plains on the
west at a height of about 5000 feet, branches off from the Yunnan
tableland between the Mekhong and Song-koi basins, and under the
general name of the *Cochin-Chinese Coast Range* is continued parallel
with the coast southwards between Annam and Siam. It throws off
several advanced spurs terminating in bold headlands on the coast,
which thus becomes disposed in a number of bays and deep inlets,

including the commodious and well-sheltered harbour of Turane. One of these spurs to the south of the Song-Koi delta serves as the natural frontier between the ancient kingdoms of Tonkin and Cochin-China. But the coast-range itself, which appears to bear no general native name, comes almost to an abrupt termination towards the frontier of Lower or French Cochin-China forming part of the Lower Mekhong **basin**. Beyond this point its further extension southwards is indicated only by some isolated eminences, such as the wooded **headland of** Cape St. James at the south-eastern extremity of the **mainland,** and further seawards by the small Pulo Condor insular **group.**

Although now inhabited chiefly by the Annamese race, *Lower Cochin-China* should more properly be called Lower Camboja. Till its conquest by the Cochin-Chinese about the middle of the last century it formed an integral part of the ancient kingdom of Camboja, and it belongs physically altogether to the same region. It comprises the greater part of the Mekhong delta, which is itself nothing more than a comparatively recent southern extension of the low-lying alluvial plains of Camboja proper. The whole region as far as and including Lake Tonlé-sap (see p. 54) constituted, probably within the last 3000 years, a marine basin penetrating far inland between the Cochin-China coast range on the east and the Pursat Hills on the west. The rapid seaward extension of the land in this direction, by which the "Great Lake" has become a land-locked basin, still inhabited by porpoises and other marine animals, is due partly to the alluvial deposits of the Mekhong, but probably still more to the gradual upheaval of the land, a movement evidently still going on, as shown by the recent conversion of the island of Cape St. James into a part of **the** mainland.

Both the delta **and still more the** inland parts **of** *Camboja* are subject to extensive **annual** floodings, which last from June to September, **and** which rise high enough to convert all the low-lying tracts into a vast inland sea, studded here and there with wooded islets serving **as** refuges for the inhabitants and their domestic animals. But after the subsidence of the waters, the whole country presents the aspect of a boundless sandy plain, diversified with numerous glittering lakes, and traversed in its entire length by the various ramifications of the Lower Mekhong fluvial system. The main stream and the Tonlé-sap emissary converging at Pnom-peñh, present capital of Camboja proper, again branch off at this point into two divergent channels, which follow a nearly parallel course through the half-submerged plains of the delta to their junction near the coast with the no less intricate system of the Donnai, or "River of Saigon." The converging point of the four great branches

of the Mekhong takes the native name of *Skådo Mukk*, that is, the
"Four Arms," or *Quatre-Bras* of the French.

Beyond the low-lying area of periodical inundations, Camboja presents
several elevated lands, chains of hills and isolated eminences concentrated
especially in the region enclosed between the Great Lake and its emissary
on the north-east and the Gulf of Siam on the south-west. These uplands,
which have not yet been systematically explored, seem to form the southern
limit of the orographic system stretching from the Yunnan tableland southwards between the Menam and Mekhong basins. They develop an irregular
series of three somewhat parallel ridges, running from the Xong hills above
Shantabun along the Gulf of Siam south-eastwards to the Mekhong delta.
Within the Siamese frontier they are known as the Pralat hills, and in
Camboja take the name of Pursat or Krevant, that is, the Cardamum
Mountains, apparently culminating in the Elephant Mountain (3600 feet),
the Pnom Popok Vil of the natives, one of the highest peaks in Camboja.

Climate.—As in other parts of the peninsula, the year is divided
into a wet and dry season, which, especially in Tonkin, succeed each
other very abruptly. Here the monsoon, which arrives in April, is
accompanied by tremendous thunderstorms, and followed by terrific
cyclones, which sweep the Gulf of Tonkin, and often cause wide-
spread ruin on the surrounding plains. In the Song-Koi delta the
moist summer heats are most oppressive even to the natives, while
the marshy fever-stricken districts are almost uninhabitable by
Europeans at this season.

Along the Cochin-Chinese seaboard the north-east monsoon sets
in about the end of October, and the south-west in the middle of
May. The heavy rains which sometimes cause the coast streams to
overflow their banks, last usually from September to December. The
glass, which in Tonkin falls as low as 45° F., seldom records less than
63° F. at Hué, capital of Cochin-China, rising in the dry season to
102° and 104° F. Yet although it never freezes or snows, the cold
is acutely felt on the elevated lands in the northern provinces above
the 15° N. lat.

In the Lower Mekhong basin (Camboja and Lower Cochin-China),
the wet season lasts from June to November, the dry from December
to May. Here the moist sultry atmosphere, especially in the Donnai
district about Saigon, is most enervating for Europeans, who after a
protracted residence become easy victims to ague, anæmia, and other
local ailments.

Natural Resources.—The advanced spurs of the Tonkin mountains, which have alone been explored, contain rich deposits of iron,
tin, copper, coal, and the precious metals, and the whole of these
highlands probably abound in mineral wealth. Gold and silver

mines also occur in the Cochin-Chinese coast range, and here the
mountain streams are washed for gold. Carboniferous limestones
prevail in the central provinces, where coal of a good quality lies in
some places close to the **surface**. In Camboja occur mines of gold,
argentiferous lead and copper, besides iron ores in great abundance,
which **have been** worked for ages by some of the semi-civilised **Kuy**
tribes in **the region to the** north-east of Lake Tonlé-sap.

Speaking of the great mineral resources of Tonkin, General W. Mesney
remarks : " That country is the centre of vast deposits of coal, in addition
to **which** the ordinary as well as the precious metals are there abundant.
I know of no less than seventeen districts in which there are flourishing
gold-fields. Silver and copper mining occupy a great deal of native and
Chinese labour. The Chinese are the proprietors of most of the larger and
more productive mines ; but the output of metal by mines surreptitiously
worked in order to avoid payment of the usual royalties is very considerable.
. . . Nearly the whole of these well-known mineral deposits are worked by
purely Chinese companies, most of the shareholders being Cantonese, many
of whom realise handsome fortunes from their mining speculations. Most
of the men employed in the mines are Kwang-si miners, who of course can
work only in their own primitive way, which reminds one of a hen scratch-
ing up the surface of the ground " (*Tongking*, Hongkong, 1884, **p. 97**).

Since the reduction of Tonkin French botanists have been study-
ing its **flora, and** the collections already forwarded by them to Paris
show that on the uplands the vegetation is of an extremely rich and
diversified character. Both here and in Cochin-China most of the
wild fauna has given place on the plains to a great variety of **cultivated** plants, such as cotton, sugar, rice, indigo, pepper, cinnamon,
pulse, and tropical fruits, besides the areca and cocoa-nut palms,
tea inferior to that of China, and tsai, a dye-wood which yields an
excellent emerald green. The forests covering the eastern slopes **of
the hills contain several** other dye-woods, teak, ebony, and **many
other valuable plants.**

Still more rich in economic products is Camboja, where the rich alluvial plains yield abundant crops of cotton, indigo, tobacco, rice, sesame,
replaced on the uplands by lac and oil-yielding trees, gutta-percha, vanilla,
cardamoms of prime quality, besides numerous cabinet and dye-woods.
Other products of this favoured region are raw silk, ivory, jerked buffalo
and elephant meat, hides and horns of the buffalo, ox, and rhinoceros ;
lastly fish, of which the Great Lake forms an inexhaustible reservoir.

Trade and Industries.—The Annamese are above all an agricultural people, devoting nearly their whole energy to the cultivation
of the soil, and leaving the pursuits of commerce and the arts mainly
in the hands of the Chinese and other foreigners. On the slopes
overlooking the plains to the usual field operations are added seri-

culture and the preparation of cotton ; but the silk and cotton fabrics woven with the raw material are much inferior to those of China. Gold and silver work displaying little taste in the designs is extensively practised, and from the Europeans in the service of the native princes during the last century some knowledge was acquired of gun-casting, engineering, and architecture.

In Annam the chief imports are cotton woven goods, salt, dried fish, tobacco, crockery, and all kinds of manufactured wares, taken in exchange for rice, opium, copper, tin, dried fungi, the edible lily, mostly for the Chinese market. To these must be added pine, largely used for coffins in China, teak, and other durable timber, eaglewood, and several other species of fragrant woods, numerous drugs, and cinnamon, regarded by the Chinese as the finest in the world. "The tusks, horns, bones, hides, and sinews of elephants, rhinoceroses, and deer are also in great demand for medicinal purposes. Honey is sent in considerable quantities to the provinces of Kwang-tung and Kwang-si, whilst a very large trade item is included under the head of various reptiles, principally snakes, which are credited with powerful medicinal properties " (Mesny, p. 49).

While the export trade with China, Siam, Singapore, and Europe is entirely monopolised by foreigners, the local traffic has been but little developed, owing partly to the indolence of the people, partly to the absence of highways of communication. In Tonkin intercourse is carried on mainly by water, and in Cochin-China there is only one highway, running from the Tonkin frontier along the coast through Hué to the Mekhong delta. But even this is kept in bad repair, and owing to the absence of bridges the rivers intersecting its track have to be forded or crossed in ferry-boats. Recently, however (1886), the French Government has invited contracts for the construction of a railway, which it is proposed to run from Hanoi for twenty-two miles to Bac-Ninh, and thence through the delta to the "Seven Pagodas."

In Camboja the chief industry is the capture and curing of fish round the shores of Lake Tonlé-sap and neighbouring waters. After supplying all the local wants, sufficient of this commodity is left for a yearly export trade, valued at nearly £300,000. Rice and fish are the staples of food amongst all the Annamese and Cambojan populations. Sericulture also forms an important branch of industry, and the Cambojans have long been noted for the excellence of their cotton and silken fabrics used in the preparation of the langutis, which form the chief feature of the national costume. Pagodas, where all the children receive their education from the bonzes, as in most other Buddhist countries, are very numerous in Camboja. But none of these modern structures can compare in size or splendour with the grand monuments of the flourishing period of Cambojan history, whose ruins are scattered over the plains and slopes stretching round the western and northern shores of Lake Tonlé-sap.

H

Political Divisions.—The whole region, which recent events have practically converted into French territory, comprises four distinct political divisions : TONKIN in the north ; COCHIN-CHINA in the centre ; LOWER COCHIN-CHINA and CAMBOJA in the south. The first two, formerly separate States, have since 1802 constituted a single kingdom, commonly spoken of as the empire of ANNAM. This term *Annam* (properly *An-nan*) appears to be a modified form of *Ngan-nan*, that is, "Southern Peace," first applied to the frontier river between China and Tonkin, and afterwards extended not only to Tonkin, but to the whole region south of that river after its conquest and pacification by China in the third century of the new era. Hence its convenient application to the same region since the union of Tonkin and Cochin-China under one dynasty, and since the transfer of the administration to France in 1883, is but a survival of the original Chinese usage, and fully justified on historic grounds.

TONKIN (Tongking, Tungking), that is, "Eastern Capital," a term originally applied to Ha-noi when that city was the royal residence, has in quite recent times been extended to the whole of the northern kingdom, whose true historic name is *Yüeh-nan*. Under the native rulers Tonkin was divided into provinces and sub-divisions bearing Chinese names, and corresponding to the administrative divisions of the Chinese empire. A native map shows the country at one time parcelled out into twenty-eight *sheng*, or provinces, each *sheng* being again grouped into a number of *hsien*, or districts. But this arrangement was subject to frequent modifications by the various Tonkinese sovereigns, and since its conquest by Cochin-China the country has been administered in much the same way as the southern kingdom. From this State Tonkin is separated partly by a spur of the coast range projecting seawards, partly by a wall built in the sixteenth century and running in the same direction. After the erection of this artificial barrier, which lies about 18° N. lat., between Hatinh and Dong-koi, the northern and southern kingdoms came to be respectively distinguished by the titles of *Dang-ngori* and *Dang-trong*, that is, "Outer" and "Inner Route."

The term COCHIN-CHINA, by which the INNER ROUTE is best known, has no more to do with *China* than it has with the Indian city of *Cochin*. It appears to be a modified form of *Kwe-Chen-Ching*, that is, the "Kingdom of Chen-Ching," the name by which this region was first known in the 9th century of the new era, from its capital *Chen-Ching*. Another although less probable derivation is from the Chinese *Co-Chen-Ching*, meaning "Old Champa," a reminiscence of the time when the Cham (Tsiam) nation was the most powerful in the peninsula, their dominion comprising the whole region between the China Sea and Menam basin. Originally Cochin-China proper reached very little south of Tonkin, with which it frequently formed one territory. But according as the Annamese race spread southwards they gradually absorbed the whole of the Cham domain as far as the extremity of the peninsula, besides a large part of the kingdom of Camboja about the Mekhong delta. Hence before the arrival of the French, Cochin-China comprised the whole of the coast lands from Tonkin nearly to the foot of the Pursat hills in South Camboja.

After the occupation of the Mekhong delta district (Lower Cochin-China) by the French in 1862-67, Cochin-China proper still comprised three main divisions : *Upper, Central,* and *Southern (Champa) Cochin-China,* which were and still are divided into nine provinces, which going southwards are as under :—

1. *Quang-binh,* separated from Tongkin by the above-mentioned Lui-Sai, or "Great Wall."
2. *Quang-tri,* a somewhat inland district.
3. *Quang-duc,* where is situated the present capital, Hué.
4. *Quang-nam,* including the fine harbour of Turane.
5. *Quang-ngai,* a rugged hilly district still partly held by wild tribes.
6. *Binh-dinh (Qui-nhon),* one of the finest and most productive provinces in the kingdom. On the coast is the port of Quinhon, one of the best on the Annamese coast.
7. *Phu-yen,* also a highly-cultivated district yielding large quantities of rice, sugar, maize, and pulse.
8. *Binh-hoa,* mostly a well-wooded upland district.
9. *Binh-thuan,* large, low-lying, thinly-peopled **district, comprising** with the foregoing the former territory of Champa.

Each of these provinces is sub-divided into two or three *phu,* and these usually into two *huyen,* which are again divided into a number of *tong,* or groups of villages and hamlets.

LOWER or FRENCH COCHIN-CHINA comprises the common Mekhong-Donnai delta, which was wrested by the Annamese from the ancient kingdom of Camboja between the years 1689—1750, and which was ceded by the Annamese to the French partly in 1862, partly in 1867. Before the cession it was divided into six provinces, which in 1876 were reduced to four administrative circumscriptions, as under :

1. *Saigon* in the east, with Saigon on the Donnai river, capital of the settlement.
2. *Mytho,* west of Saigon, as far as the eastern mouth of the Mekhong.
3. *Vinh-long,* about both mouths of the Mekhong.
4. *Bassac,* thence to the Cambojan frontier.

H 2

CAMBOJA. At one time embracing a great part of Further India between the China Sea and Gulf of Siam, the ancient and flourishing empire of Camboja has been gradually reduced by Siamese encroachments on the north and west, and by Annamese conquests on the south and east to a territory of scarcely 40,000 square miles on both sides of the Great Lake and its emissary, with a contracted seaboard of some 80 miles between the Mekhong delta and Siam. To save this remnant of his possessions from complete absorption by those rival States, the feeble descendant of the royal builders of Ankhor-Vat was fain to accept the proffered protectorate of France in 1864. But this protectorate was practically converted into annexation by the treaty of 1886, in virtue of which the king constituted himself a pensioner of France, abdicating all his royal functions, and handing over the administration of the kingdom to the Governor of Saigon.

As constituted at the date of this treaty, Camboja comprised the five subjoined *deys*, or historic divisions, which possessed no administrative character, but served rather as appanages for the five ministers composing the Royal Privy Council :

1. *Compong-Srui*, north of the Great Lake and its emissary.
2. *Treang*, between the western branch of the Lower Mekhong and the coast.
3. *Thang-Khmun*, on the Mekhong.
4. *Ba-Puom*, east of the Mekhong.
5. *Pursat*, south of the Great Lake.

These appanages were divided into fifty-six *khet*, or provinces, each administered by a governor who was appointed by the king, and who generally selected his own lieutenants, sub-prefects, and other minor officials.

Administration.—The Annamese government has hitherto been an absolute despotism, tempered only by a few traditional formulas, and recently by treaties with France. Finance, war, justice, home affairs, religion, and public works form so many ministries, with a president, two vice-presidents, two councillors, and a vice-chancellor. But over all, overshadowing the throne itself, stand the Chief-Censor, head of the lettered classes, assisted by two or three vice-censors, for civil affairs ; and for war the marshall-in-chief, assisted by four marshalls, all these dignitaries forming the "Columns of State."

Notwithstanding the absolute character of the central Government, Annam enjoys extensive communal rights, and here, as in China, there is no hereditary aristocracy, all imperial functions being reserved for successful candidates at the public academy, corresponding in some respects to our Civil Service examinations. The whole population is further grouped in two main divisions : the *inscribed*,

including all paying a poll-tax, and whose names figure in the sche-
dules of taxation; and the *non-inscribed*, including day labourers,
the poor, and all living from hand to mouth. But the native administration appears to have been practically set aside in 1886, when a
Resident General with almost absolute jurisdiction was appointed to
control the civil and military affairs of Annam. The first person
chosen to fill this high office was the distinguished French *savant*,
M. Paul Bert, member of the Academy of Sciences. Soon after his
arrival M. Bert issued a proclamation to the people, assuring them of
his intention to afford them every protection in his power, but giving
them indirectly to understand that the management of affairs had
been transferred from the native to the French authorities. M. Bert
fell a victim to the treacherous climate in November 1886.

The social organisation of the people into two classes has been maintained in French Cochin-China, which is under the administration of a
governor assisted by a Privy Council. Since the treaty of 1886, Camboja has been brought within the jurisdiction of the French governor,
whose residence is at Saigon. The king of that country, now a pensioner
of France, has finally surrendered all his regal functions, and at his death
Camboja will probably be merged with Lower Cochin-China in one colonial
settlement, both districts forming part of the same geographical area.

Topography.—There are but few large towns in Annam, where
the agricultural population is somewhat evenly distributed in rural
villages and hamlets thickly scattered over all the arable lands.
Here the northern capital *Hanoi* (*Ho-nei*), first known as *Tonkin*,
or the "Eastern Capital," stands on the main navigable branch of
the Song-koi, near the head of the delta, and about 100 miles from
the coast. Hanoi, also called *Kesho*, was first opened to the trade of
the world by a treaty concluded with France in 1874. But the
expectations that through it an important trade route might be
opened to Yunnan and Southern China along the Song-Kai seem
doomed to disappointment, owing to the numerous rapids obstructing
the main head waters of that river. Hanoi is well built with brick
or stone houses, marble pavements, and a vast citadel nearly four
miles in circumference, built by French engineers at the close of the
last century. The proper outport of Hanoi is *Haiphong* (*Haipong*),
also opened to European trade in 1874. It stands at the mouth of a
northern branch of the Song-koi delta, communicating by a navigable
channel with the populous and strongly fortified town of *Haidzuong*
in the Thai-binh delta. Near the Chinese frontier lies the important
strategical post of *Langson* (*Lang-shin*), memorable for a signal defeat

of the French **during** the recent military operations against the
"Black Flags."

Hué (*Thua-Thien*), present capital of Cochin-China and of the
whole of Annam, lies on the coast some miles north-west of the Bay
of *Turane*, and, like Hanoi, boasts of a vast citadel constructed by
French engineers early in this century, and comprising barracks,
arsenals, granaries, and artillery grounds. On a neighbouring hill
stand the glittering domes of the royal mausoleum, containing the
tombs of many kings lavishly ornamented with gems and the precious metals. Yet in the fourteenth century Hué was still spoken
of as comprised within the limits of the Champa State, showing how
very recent has been the spread of the Annamese race south of its
true home in Tonkin. It is connected northwards with Hanoi, and
southwards through Tourane with Saigon by a postal service along
the imperial highway, with stages at intervals of ten to twelve miles.

Saigon, at the confluence of the Saigon and Donnai rivers a few
miles from the coast, has almost assumed a European aspect since
its occupation by the French in 1859. Although situated in an
extremely hot and unhealthy district, its favourable position on
navigable waters, communicating by a deep channel with the fortified
port of *Vinh-long* and *Mytho* on the eastern branch of the Mekhong
delta, has caused it to be chosen as the seat of government for all the
French possessions in the Donnai-Mekhong basin. *Chandoc*, the
chief place on the western branch of the Mekhong within French
Cochin-China, has also the advantage of water communication with
the coast by a navigable canal running southwards to *Hatien* on the
Gulf of Siam near the Cambojan frontier.

Beyond this frontier stands *Kampot*, the only seaport in Camboja
proper, with a deep and well-sheltered harbour, **which** has been
frequented for many ages by Malay and Chinese traders. Within
the last few years the seat of the Cambojan government **has** been
transferred from *Udong* on the Tonlé-sap emissary to *Pnom-penh*, a
little farther down, at the meeting of the "Four Waters." But the
still older capital, *Angkor*, which stood at the north-western margin
of the Great Lake, and which appears to have flourished when that
basin was still the head of a marine inlet, has long been a city of the
dead, whose glorious past is attested only by the ruins of its stupendous monuments. For the ancient Cambojan culture "seems to **have**
subsided with the subsidence of the waters. The Great Lake with-
drew from Angkor, the marine inlet became gradually filled in, the
surrounding plains were converted into marshes, the population

melted slowly away, and to the busy cities and thronged temples succeeded the scattered hovels of a decrepit people, who have lost the very traditions of a glorious past" (Keane's *Reclus*, viii. p. 484). They have even surrendered the very site of these monuments. A lacustrine district profusely strewn with the vestiges of cities, temples, and fortresses, all included within the Siamese frontier by the treaty of 1864, determining the extremely conventional limits of the conterminous States.

Historical Notes.—Mention occurs of the Annamese people under the name of *Giao-shi* (Kiao-shi, Chiao-chih), and of the country under that of *Yueh-nan* in the Chinese legendary records as early as 2285 B.C. Subsequently it was known by various other names, such as Lo-yueh-ti, Nan-chiao, Lu-liang, &c., until in 1175 A.D. *An-nam* (properly Au-nan) became the official name of the whole region by order of the Emperor Hsiao-tsung of the southern Sung dynasty. The terms *Tonkin* and *Cochin-China* in their modern acceptation are of comparatively quite recent origin, just as the severance of the country itself into two distinct States is a comparatively recent historic event. It may be assumed that the Annamese domain was originally restricted to the region of the Song-koi delta, where it was encircled on the north by China, on the west by the Moi and Muong hill tribes, on the south by the Champa (Tsiampa) territory. Its historic growth took place almost entirely in a southerly direction along the strip of low-lying coast-lands between the coast range and the China Sea, where it gradually encroached upon and finally absorbed the whole of the Champa domain. Then the coast-lands became constituted into a separate State distinct from and independent of the northern kingdom, an artificial barrier being constructed between them in the sixteenth century, and separate names, such as Tonkin and Cochin-China, coming gradually into use to distinguish the two Annamese kingdoms.

From the remotest times China claimed, and intermittently exercised, suzerain authority over Annam, whose energies have for ages been wasted partly in vain efforts to resist this claim, partly in still more disastrous warfare between the two rival States. Almost the first distinctly historic event was the reduction of Lu-liang, as Tonkin was then called, by the Chinese in 218 B.C., when the country was divided into prefectures, and a civil and military organisation established on the Chinese model. On this occasion a large number of Chinese emigrants are said to have settled in the country, where they amalgamated or became gradually assimilated with the aboriginal Giao-shi stock, leaving, however, traces of their influence perceptible to the present day in the mixed character of the Annamese current speech.

Early in the ninth century of the new era the term Kwe-Chen-Ching (Cochin-China) began to be applied to the southern, which had already asserted its independence of the northern, kingdom. In 1428 the two States freed themselves temporarily from the Chinese protectorate, and 200 years later the Annamese reduced all that remained of the Champa territory, driving the natives to the uplands, and settling in the plains. This conquest was followed about 1750 by that of the southern or maritime provinces of Camboja since known as Lower (now French) Cochin-China.

In 1775 the King of Cochin-China, who had usurped the throne in 1774, reduced Tonkin, and was acknowledged sovereign of Annam by the Chinese emperor. But in 1798 Gia-long, son of the deposed monarch, recovers the throne with the aid of some French auxiliaries, and in 1802 reconstitutes the Annamese empire under the Cochin-Chinese sceptre. From this time the relations with France become more frequent, and French preponderance is firmly established when through the efforts of Pigneau de Béhaine, Bishop of Adran, effect is given to the treaty of 1787, ceding to France the Bay of Turane and the Pulo Condor islets, in return for her assistance in restoring the old Cochin-Chinese dynasty under Gia-long. French officers are now employed in drilling the native troops, fortifying the royal palaces, erecting formidable citadels at Hué, Hanoi, and other important strategic points, and affording the restored king the moral force by which he triumphs over all his adversaries.

But after his death in 1820 the anti-European national party acquires the ascendant, the French officers are dismissed, and the Roman Catholic religion, which had made rapid progress during the reign of Gia-long, is subjected to cruel and systematic persecution. Notwithstanding the protests and occasional intervention of France, this policy is persevered in, until the execution of Bishop Diaz in 1857 by order of Tu-Duc, third in succession from Gia-long, calls for more active interference. Admiral Rigault de Genouilly captures Tourane in 1858, followed next year by the rout of the Annamese army at the same place, and the occupation of the forts at the entrance of the Donnai and of Gia-dihh (Saigon), capital of Lower Cochin-China. This virtually established French supremacy, which was sealed by the treaty of 1862, ceding the three best, and that of 1867 the three remaining provinces of Lower Cochin-China. It was further strengthened and extended by the treaty of 1863, securing the protectorate of Camboja and the important strategical position of " Quatre-Bras " on the Mekhong.

Then came the scientific expedition of the Mekhong (1866-68), which dissipated the hopes entertained of that river giving access to the trade of Southern China. Attention was accordingly now attracted to the Song-koi basin, and the establishment of French interests in Tonkin secured by the treaties of peace and commerce concluded with the Annamese Government in 1874. This prepared the way for the recent diplomatic complications with Annam and China, followed by the military operations in Cochin-China and Tonkin, which led up to the treaties of 1883 and 1884, extending the French protectorate to the whole of Annam, and forbidding the Annamese Government all diplomatic relations with foreign powers, China included, except through the intermediary of France. Lastly, the appointment in 1886 of a French Resident-General, with full administrative powers, effaced the last vestige of national autonomy, and virtually reduced the ancient kingdoms of Tonkin and Cochin-China to the position of an outlying French possession. But here again the commercial speculation was doomed to disappointment, the recent exploration of the Song-koi and its head waters having shown that it is as little suited as the Mekhong for opening up the trade of Southern China to the political masters of Annam.

On the other hand, the French authorities appear to have already succeeded in stopping the wholesale massacres of Christians which had broken out afresh during the late international troubles. Christianity, which was originally introduced by the Portuguese Jesuits in the seventeenth century, and which had spread rapidly during the early years of the present century, will now receive a fresh impetus under the French missionaries, who have organised five vicariates under the *Missions Étrangères* of Paris, and three under the Dominicans. The movement is favoured by the anti-religious French officials, who are aware that in the East the native Roman Catholic is first a Frenchman and then a patriot.

STATISTICS OF INDO-CHINA.

AREAS AND POPULATIONS.

	Area in Sq. Miles.	Approx. Pop.
British Indo-China (Burmah)	300,000	8,000,000
Native Indo-China (Siam)	290,000	25,000,000 (?)
French Indo-China (Annam and Camboja)	223,000	20,000,000 (?)
Total Indo-China	813,000	53,000,000

APPROXIMATE POPULATION OF INDO-CHINA ACCORDING TO RACE.

Burmese, Talaings, and Karens	7,000,000
Annamese (Tonkinese and Cochin-Chinese)	19,000,000
Siamese, Shans, and Laos	17,000,000
Chinese	5,000,000
Cambojans	1,700,000
Kachyens, Mois, and other wild tribes	750,000
Malays and Chams	50,000
Total	50,500,000

BURMAH.

	Area in Sq. Miles.	Population.
Upper (late Independent) Burmah	210,000 (?)	4,000,000
British Burmah (1881) Arakan	14,500	588,000
Pegu and Irawaddy	26,000	2,323,000
Tenasserim	47,000	826,000
Total	297,500	7,737,000

CHIEF TOWNS OF BURMAH.

	Population.			Population.
Rangûn	134,000		Thayetmyo	12,000
Mandalay	70,000		Allanmyo	11,000
Maulmain	54,000		Mergui	11,000
Prome	30,000		Kyau-ghin	10,000

	Population.			Population.
Bassein .	25,000	Yandûn .	.	10,000
Akyab .	25,000	Magwé .	.	8,000
Henzada .	18,000	Moné .	.	8,000
Meng-gyi .	17,000	Shwegyeng	.	8,000
Myi Kyan	15,000	Pegu .	.	7,000
Tavoy .	15,000	Bhamô .	.	6,000
Shwe-doung	14,000	Muchabo .	.	5,000
Toung-ngu	14,000			

Exports, Upper to Lower Burmah (1884) . . £1,706,000
Imports ,, from ,, ,, ,, . . 1,826,000
Revenue, Upper Burmah ,, . . 80,000
Expenditure ,, ,, . . 40,000

SIAM.

CHIEF TOWNS.

	Population.			Population.
Bangkok .	600,000	Nakhon-Savan		12,000
Ayuthia .	50,000	Petrin .	.	10,000
Zimmé .	50,000	Meklong	.	10,000
Lakhon .	25,000	Paklat .	.	7,000
Hluang-Prabang	15,000	Korat .	.	7,000
Pré . .	15,000	Paknam .	.	7,000
Labong	12,000			

Bangkok, Imports (1884) £1,145,000
,, Exports ,, . . . 2,052,000
Total average exports of Siam . . . 6,000,000

1880. Exports to Great Britain, £341,000 ; Imports £23,000
1884. ,, ,, ,, 162,000 ; ,, 44,000
1884. Bangkok, shipping : 418 vessels of 241 tons,
of which British 240 ,, 151 ,,
Mercantile Marine : 60 vessels of 21,000 tons.
Navy : 14 steamers of 5,815 tons, and 51 guns.

Average Revenue of Siam £3,000,000

ANNAM AND CAMBOJA.

	Area in Sq. Miles.	Population.*
Tonkin . . .	120,000 (?)	12,000,000 (?)
Cochin-China .	46,000	5,000,000 (?)
Lower Cochin-China	24,000	1,900,000
Camboja . .	33,000	1,020,000
Total	223,000	19,920,000

* The population of Tonkin, usually given at 15 and even 20,000,000, appears
certainly not to exceed 12,000,000. Bouinnais and Paulus (L'Indo-Chine française,

CHIEF TOWNS.

	Population.			Population.
Hanoi (Kesho)	130,000	Mytho	.	15,000
Saigon	65,000	Udong	.	12,000
Hué	60,000	Hakoi	.	10,000
Cholon	50,000	Song-tai	.	8,000
Namdiñh	50,000	Turane	.	3,000
Haidzuong	50,000	Kampot	.	2,000
Pnom-Peñh	30,000			

POPULATION ACCORDING TO RACES.

	Lower Cochin-China.	Camboja.
Annamese	1,710,000	5,000
Cambojans	110,000	600,000
Chinese	60,000	380,000
Chams and Malays	10,000	30,000
Hill-men and others	10,000	5,000
Total	1,900,000	1,020,000

Saigon, Imports, average £2,750,000
  ,, Exports, ,, ... 3,500,000
  ,, Shipping : 850 vessels of 700,000 tons
  ,, British Shipping : 336 vessels of 284,750 tons

Haipong, Imports (1880), £218,000 ; Exports, £300,000
  ,, Shipping (1880) : 458 vessels of 124,000 tons
  ,, British Shipping (1880) : 35 per cent.

Hanoi, Trade with Yunnan by the Red River (1880) **£140,000**

Lower Cochin China, Revenue (1882) £ 875,000
  ,, ,, ,, Expenditure ,, 1,100,000
Camboja, Average Revenue . . . 130,000
  ,, ,, Exports . . . 200,000

1885) estimate it roughly at from 9 to 12,000,000, adding that an exaggerated idea was formed of its density from the fact that the delta is thickly settled, while all the rest of the country is very thinly peopled. J. G. Scott also states that the population is probably not more than 12,000,000.—(Proc. Geo. Soc., April, 1886.)

PART III.

THE EASTERN ARCHIPELAGO.

CHAPTER I.

General Survey.—Like the other great continental masses in
the northern hemisphere, Asia is continued seawards at its south-
eastern extremity by a vast insular region, which is variously known
as the EASTERN, the ASIATIC, the MALAY, and even the EAST INDIAN
ARCHIPELAGO. Although now too firmly established to be conveniently set aside, none of these expressions can be accepted as entirely
adequate, being either too vague, or else implying half truths, or
even suggesting erroneous views. Thus, while "East Indian" can
scarcely be justified at all, it will be seen that "Asiatic" and "Malay"
are applicable only to one section of this oceanic world; so that,
notwithstanding its somewhat indefinite character, the title here
adopted seems on the whole the least objectionable.

It was formerly assumed that the Eastern Archipelago constituted
a homogeneous physical region, forming a natural connecting link
between Asia and Australia, or rather representing the remnants of
a continuous tract by which those continents were at one time united.
But George W. Earl and Alfred Russel Wallace, basing their induc-
tions on a more accurate knowledge of the oceanic depths and of the
geology and natural history of the islands themselves, have clearly
shown that they comprise at least two main divisions, a Western, or
"Indo-Malayan," and an Eastern, or "Austro-Malayan," which may
be safely regarded as respectively forming a southern extension of
the Asiatic, and a northern extension of the Australian continent.
A ship sailing from the head of the Gulf of Siam southwards will

traverse extremely shallow waters, scarcely anywhere exceeding 40 fathoms all the way to Bali, at the eastern extremity of Java. But here the sounding line plunges suddenly into great depths, which, beginning with the narrow passage 15 miles wide between Bali and Lombok, are continued northwards and eastwards through Macassar Strait and the Celébes Sea to the Pacific Ocean. Hence the dividing line is drawn in the same direction between Bali, Borneo, and **the** Philippines **on the one hand, and** Lombok and Celébes on the other, all the land to the left **or west** of this line being assigned **to** Asia, all the rest to Australia.

As the western (Wallace's "Indo-Malayan") division lies mainly on a shallow sub-marine plateau, which seldom exceeds 40, and nowhere reaches more than 100 fathoms, and as its flora and fauna also correspond **on** the whole to those of southern Asia, it seems in every way reasonable to regard this division as belonging to the Asiatic mainland, whether detached from it by subsidence, or loosely attached to it by upheaval in post-miocene times. It comprises the Philippines and the Greater Sunda Islands, that is, Borneo, Sumatra, and Java, with all the adjacent isles, together with Bali, westernmost of the Lesser Sunda Islands, or considerably more than half of the whole region.

But there are weighty reasons which militate against the view that assigns the eastern (Wallace's "Austro-Malayan") division indiscriminately to Australia. Here many of the islands, notably Celébes and its numerous dependencies to the east, lie not on a submarine plateau in shallow waters, but in a very deep inner basin, ranging from some 500 to 4000 fathoms and upwards. In fact, the Celébes and Banda Seas, north and south of Celébes, are by far the deepest of all these inland waters, jointly constituting a profound abyss, in some parts over 700 miles wide, which flows from the Indian to the North Pacific Ocean, between the Asiatic and Australian submarine plateaux west and east. It stretches from Bali eastwards to Timor, and even to Timor-laut, so that the 100 fathom line indicating the extreme limits of the Australian submarine plateau runs at a comparatively short distance from the mainland in this direction. Further east, however, the Arafura Sea, between Australia and New Guinea, comes entirely within these limits, and as their fauna and flora also largely correspond, New Guinea and its dependencies must in any case be assigned to the Australian world. In the same division may further be conveniently included all the Lesser Sunda Islands except Bali, for although washed by deep seas, this volcanic and no doubt comparatively recent group has received most of its animal and vegetable species not from the neighbouring Asiatic, but from the more remote Australian section. In the Austro-Malayan division will therefore be comprised New Guinea, with the adjacent western islands of Waigiou, Salwatty, Mysol, Aru, &c.; and of the Lesser Sunda group, Lombok, Sumbawa, Floris, Chandana, Timor, Timor-laut, and intervening islets.

There remains the great island of Celébes, with its immediate and more remote eastern dependencies, Muna, Bouton, Sula, Buru, Ceram, Jilolo, Bachian, &c., that is, the Molucca and Banda groups. These occupy the very centre of the Eastern Archipelago midway between the Asiatic and

Australian worlds, from which they are severed not only by some of the deepest waters on the globe, but also by a fauna, and to a large extent by a flora, presenting a surprising number of absolutely independent forms. Here it will suffice to mention the spices (nutmeg, clove, &c.) of the Banda Isles, the Babirusa, or Hog-deer, the Sapi-utan (Anoa depressicornis), half ox, half antelope, and the curious maleo, or brush turkey, all peculiar to Celèbes and adjacent isles. When it is added that Celèbes forms the eastern limit of range for the squirrel and lemur, and the western for the cuscus, or eastern opossum of the marsupial order, and further that its geological formation appears to be far older than that of the surrounding volcanic Sundanese groups, the inference seems irresistible that these islands form a separate oceanic division independent alike of the Asiatic and Australian worlds. They appear to be the last eastern fragments of a vanished miocene continent, to which Sclater has given the name of Lemuria, and whose furthest western extension is indicated by the great island of Madagascar. Here have survived certain early organic types, which were doubtless at one time diffused over a far wider range throughout a miocene continent, whose subsidence has made room for the more recently-appeared volcanic formations in the Eastern Archipelago. The comparatively modern appearance of these volcanic lands has been noticed by all careful observers ; amongst others, by H. O. Forbes, who speaks of fossil plants and of shells (Ostrea and Pecten) in West Java closely resembling those in the adjacent seas, "and showing that an elevation of some 200 to 300 feet had taken place here at a recent period."—*A Naturalist's Wanderings in the Eastern Archipelago*, p. 63.

Volcanic Formations.—These constitute one of the largest and most active igneous regions in the world, sweeping in a continuous chain from near the northern extremity of Sumatra, through Java and all the Lesser Sunda group, eastwards to Timor and Wetter, thence curving round to Nila (130° E. long.), and back to Buru, and thence northwards to Tidor, Ternate, and Jilolo. Here the volcanic belt shifts suddenly westwards to the northern extremity of Celèbes, whence it is continued northwards through Siao, Sanguir, and the Philippines to the north end of Luzon. Yet this chain is itself but a link in a still vaster system, which, through Formosa, Japan, Kamchatka, the Aleutian Islands, and west coast of America, encircles the whole of the Pacific, and which may be traced at intervals round the Indian Ocean through Barren Island (Andaman), Ramri, and Cheduba on the Arabian coast, the submarine volcano off the Coromandel seaboard, Kenia, Kilima-Njaro, and the Komoro Islands in East Africa, and neighbouring waters.

In the Eastern Archipelago many of the burning mountains attain considerable altitudes. In the Philippines and North Celèbes none appear to rise higher than 6000 or 7000 feet ; but in Java and Sumatra several have an elevation of over 10,000, while two even exceed 12,000 feet. These, however, are not the culminating points of the whole Archipelago, as is generally supposed, for the granitic Kini Balu in North Borneo considerably exceeds 13,500 feet. But by far the highest summits are found in New Guinea, where some of the Owen Stanley peaks reach 13,000, and those of the Charles Louis Mountains 17,000 or 18,000 feet above sea-level. This is the highest land which occurs anywhere between the Himalayas in the west, and the Cordilleras de los Andes in the east.

Throughout the whole of the volcanic belt in the Eastern Archipelago, which has a total length of not less than 5000 miles, and which contains some sixty active and hundreds of extinct volcanoes, earthquakes of varying intensity are still of almost constant occurrence. These are at times accompanied by tremendous eruptions, causing wide-spread ruin over vast spaces, and changing the very aspect of the land. The most recent, and one of the most memorable of these outbursts, occurred on August 26, 1883, when the island of Krakatao, in the Sunda Strait between Java and Sumatra, was almost blown to pieces, both sides of the strait wasted far and wide, the surrounding waters strewn with floating lavas for hundreds of miles, and the atmosphere filled with such a prodigious quantity of impalpable dust, that to it were attributed the remarkable crepuscular lights visible in almost every part of the world for months afterwards. Although since this event the Javanese volcanoes have been comparatively quiet, Smeroe, the highest in the island, and its neighbours, Bromo and Lamonyon, show constant signs of disturbance. In 1885 Smeroe overwhelmed plantations and villages with eruptive matter, and lavas are continually flowing from Merapi, in the centre of Java.—Van Geuns.

Geology.—Throughout the whole of the northern section of the Archipelago, from Sumatra to the Philippines, the salient geological features seem to resemble those of the Malayan Peninsula, where an elevated granite axis is flanked at the base by palæozoic schists and slates underlying detached masses of crystalline and other limestones. The main axis of Sumatra, running in the same direction, appears to be also granitic, if not stanniferous, like the neighbouring islands of Biliton, Bintang, and Banca. In South Sumatra Forbes found eocene tertiary rocks underlying more recent pumicestone tuffa. Granitic insular groups, such as the Natunas, are also thickly strewn in the "sea of the twelve thousand islands" flowing between Sumatra and Borneo. The latter island presents the first extensive development of stratified rocks, carbonaceous of various ages, brown and yellow sandstones and shales, with intercalated grits and conglomerates, and occasional granitic outliers. Coalfields, some evidently much older than those of Labuan, and allied, perhaps, to the mesozoic carboniferous formations of East Australia, are widely diffused throughout Borneo, where gold and tin also occur in the Kinebetangen basin, and diamonds in the Landak district on the west coast.

Although Luzon, Mindoro, and some of the other larger islands in the Philippines are mainly volcanic, there are also some stratified areas with coral and other marine fossils, probably of miocene and pliocene age. Gold is found in Mindanao, which also contains limestones and eocene rocks. In the Calamiane group, between Mindoro and Palawan, limestones akin to those of the Malay Peninsula are found associated with more recent eruptive rocks. Here the island of Coron " presents to the sea a magnificent rampart of limestone cliffs and pinnacles from 600 to 1500 feet. The aspect is grandly picturesque, the bluish-gray rocks, with patches of brilliant red, yellow, pink, dark and light green, descending precipitously into forty fathoms of water. The outline of the island is magnificently rugged and irregular, weathered into needles and pinnacles of the most fantastic shape, in the recesses of which there is much pale green grass, with patches of darker jungle."—Tenison-Woods. In general the substratum of the Philippines appears to be formed of crystalline schists, above which rise extensive eruptive rocks of an archaic type, in many places subsequently modified by the action of sulphurous emanations.— A. F. Renard.

New Guinea, like Borneo, occupies a quiescent area apparently quite free from active if not from extinct volcanoes. Crystalline rocks, however, occur on the north coast, and the Arfak range behind Geelvink Bay seems to be mainly granitic. But elsewhere stratified formations are the prevailing feature. Horizontal sandstones weathered into fantastic forms cover many of the islands along the south-east coast, where some limestone hills are found underlying auriferous quartzose sands.

Gold, as well as iron, tin, and copper, is also found in several parts of Celebes, where granitic rocks no doubt largely prevail. But the extreme southern peninsula is traversed by basalt and limestone ranges, while the extreme north-eastern point is igneous, and occupied by several active volcanoes. The clusters of islands at the extremities of the peninsulas, both in Celebes and Jilolo, evidently at one time formed part of the mainland, so that the peculiar form of these islands—a central nucleus, with four limbs radiating to the north, south, and east—is probably due to subsidence of the eastern valleys, now replaced by deep marine inlets.

Some of the rivers are also washed for gold in Timor, where crystalline limestones are found interspersed with rough grey crystalline sandstones, the latter in some places cropping out or embedded in horizontal masses of sand and small gravel. This black shingly detritus points to a not very remote time when the whole of East Timor formed the bed of a marine or lacustrine basin, which was afterwards slowly upheaved, and may still be rising.—H. O. Forbes. From this point westwards to Java all the Lesser Sunda Islands, except, perhaps, Rotti and Sumba (Sandalwood), are essentially volcanic, and also of comparatively recent origin.

Extent—Population.—The Eastern Archipelago is by far the largest insular group in the world. It contains two islands, Borneo and New Guinea, each larger than the British Isles and France taken

together; three others, Celébes, **Java**, and Luzon, all as large **as**
Ireland; another, Sumatra, about equal to Great Britain; "eighteen
more on an average as large as Jamaica; more than a hundred **as**
large as the Isle of Wight; isles and islets of smaller size innumerable."—Wallace. Including land and water, it covers a space far
larger than Europe, while the absolute extent of land cannot be
estimated at much less than 1,300,000 square miles. Stretching from
the westernmost point of Sumatra across fifty-six degrees of the
meridian (95°—151° E. long.), eastwards to the furthest extremity of
New Guinea, and from the north end of the Philippines across nearly
thirty degrees of latitude (18° N.—11° S.) southwards to Roti, south
of Timor, it has a total length of about 4000 miles from west to east,
and an extreme breadth of over 2000 from north to south. The
population, consisting mainly of Indonesian, Malay, and Papuan
elements, and roughly estimated at some 34,000,000, is most unequally distributed, considerably more than half being concentrated
in the rich and highly cultivated island of Java, while of the
remainder rather more than one half are centred in the northern
Philippine group. For the respective areas and populations of the
three great divisions—*Asiatic*, or *Western*, *Oceanic*, **or** *Central*, and
Australian, **or** *Eastern*—the reader is referred to the Statistical
Tables.

CHAPTER II.

CLIMATE—FLORA—FAUNA.

Climate.—Intersected by the equator, about which most of the
land is disposed in nearly equal proportions between 10° N. and 10°
S. lat., the Eastern Archipelago enjoys, like all equatorial regions, **an**
essentially tropical climate characterised by great heat and moisture.
Owing to the generally high temperature of the surrounding waters,
and the regular recurrence of the periodical wet north-west monsoons,
these elements prevail more uniformly in this oceanic world than in
any other part of the globe. But the south-east monsoon, which
comes from the arid plains of Australia, and lasts from March to
November, is necessarily of a somewhat dry character. Hence the
islands exposed to its influence, that is, the Lesser Sundanese

group, with the eastern extremity of Java, are drier than other parts of the Archipelago. That this contrast is due to the influence of the Australian continent appears evident from the fact, that here the rainfall increases in abundance according as we proceed from Timor westwards to Java, the greater part of which island comes **within** the range of the wet monsoons from the Indian Ocean. East of **Java** and Bali very little moisture is precipitated during the prevalence of the south-eastern trade winds, and "towards the latter end **of this dry** season the drought is so great that many streams dry **up, and most of** the trees lose their leaves. The heat is then intense ; **and** were it not that the nights **are** cool and the breeze always blowing, the climate would approach in severity that of Australia itself." —Wallace.

In Sumatra also great heats prevail, especially on the extensive open plains, such as that of Pertibi in the Batta country, which are exposed for months together to dry scorching winds, raising the temperature to 95° or 97° in the shade. But in Java the glass seldom rises above 90° F. even at Batavia, while on the uplands from 3000 to 5000 feet it ranges from 50° to little over 70° throughout the year. Java, however, suffers at times from long droughts, followed by excessive rains, causing disastrous inundations in one part of the country, while other places are suffering from an absolute want **of** water. This anomaly is attributed to the monsoons, which blow irregularly, and which cause more anxiety to the Javanese than their ever restless volcanoes.—Van Geuns. At Manilla in the Philippines, **with a heavy** rainfall of 98 or 100 inches, the variations of temperature **are limited** to 72° and 95° F., the greatest heats occurring in the **months between** April and August. The northern parts of these **islands are exposed** to the south-west and to the still more violent north-east monsoons, the changes in the direction of these winds being accompanied by terrific typhoons, which are most dreaded in October, but which never reach further south than about 10° N. lat. Hence the numerous inner seas separating the various secondary archipelagoes—the Celebes Sea, between Mindanao, Borneo, and Celebes, the Banda Sea, between Ceram and Timor, the Java Sea, between Java and Borneo, the Arafura Sea, between New Guinea and Australia—are mostly still-water basins, freely navigated in their open praus by the Malays, Bugis, Sundanese, and other seafaring populations.

The climate of North Borneo is also described by Dr. Walker as remarkable especially for its equable character and the absence of extremes.

The temperature, rainfall, winds, natural phenomena generally, and the diseases, are, for a tropical country, of the most mild and temperate types. The country is visited by the regular monsoons at the ordinary times; the rainfall near the coast ranging from 156 to 101, and averaging 124 inches, and the temperature lying between 67° and 94° F. As might be expected, there are neither typhoons nor earthquakes, the only present indication of volcanic action being a hot spring reported to exist in an islet off the coast.—*North Borneo Herald*, July, 1886.

Flora and Fauna.—Thanks to its position in the midst **of a vast** sea heated by the tropical sun, the Eastern Archipelago presents almost everywhere the aspect of a forest region overgrown with a rich and varied vegetation, from sea-level to the summits of its highest mountain ranges. This is mainly true of Sumatra, Borneo, the Philippines, the Moluccas, and New Guinea, as well as Java and Celébes, in all their unreclaimed districts. The chief exceptions are Timor and the Lesser Sunda group, where forest tracts are rather the exception than the rule, a contrast sufficiently explained by the proximity of the Australian mainland, and the hot, dry south-east winds blowing **from** that region for the greater part of the year.

Another and more striking contrast is that presented by the vegetable forms respectively characteristic of the Asiatic and Australian divisions of the Archipelago. Here all resemblance and analogy cease abruptly, a narrow marine channel being sufficient to separate the two organic worlds in some places, and notably between Bali and Lombok, where the Asiatic sub-marine tableland suddenly ceases.

Nor is the contrast limited to the vegetation, but also extends to **the** animal kingdom, and even in some respects to man himself. The elephant and tapir of Sumatra and Borneo, the rhinoceros of Borneo, and the allied Javanese species are also found in the neighbouring Asiatic lands, pointing to a time when these great islands still formed part of the continent. These analogies, which extend **to** birds and insects, may also be traced as far as the Philippines, although here longer isolation has greatly diminished resemblances and intensified divergencies.

But when we pass **over** to the Australian **division all is different,** and the contrasts become more marked at **every** step. Here **no** elephants, no members **of** the canine and feline groups, no urangs, gibbons, or other apes, no deer, sheep, or oxen, in a word, no large mammals of any sort; but **in** their place the lower mammals and marsupials, or pouched animals, of which the kangaroo is typical. Here also the loric and flying fox, and still more curious ornitho-

rhynchus, half bird half mammal, all common to New Guinea **and**
neighbouring islands, and extending through Timor and the Lesser
Sundanese isles to Bali, where the Australian fauna ceases **and the
Asiatic begins.**

As already remarked by Wallace, these great contrasts are entirely
independent of climate, which is nearly the same everywhere. The
sudden changes in the organic world must be traced back to former changes
in the distribution of land and water, for they take place without any
corresponding modifications of the present environment. They are even
independent of the volcanic belt, which strikes across both sections without
determining any appreciable differences in their living forms. Borneo and
New Guinea, again, both belong to quiescent or non-volcanic areas, and
both are exposed to the same climatic conditions. Yet the contrast
between their animal and vegetable species is extreme.

Fresh contrasts in the Oceanic division of the Archipelago, where the
plants, and still more the fauna, present numerous types absolutely distinct from those both of the Asiatic and Australian divisions. Reference
has already been made to the spices of the Moluccas, which, however,
appear at some remote period to have invaded the Indo-Chinese Peninsula.
But there are no counterparts anywhere to be found to the Babirusa, Sapi-
utan, and Maleo of Celebes. In the same island Wallace finds 80 out of
128 species of birds, 11 out of 14 terrestrial mammals, 86 out of 118
butterflies, and a very large number of beetles quite peculiar, and occurring
nowhere else in the archipelago. "The student of geographical distribution," observes this distinguished naturalist, "must see in the extraordinary and isolated productions of Celebes proofs of the former existence
of some continent, whence the ancestors of these creatures and of many
other intermediate forms could have been derived."—*Malay Archipelago*,
ch. xviii.

CHAPTER III.

INHABITANTS—MALAYS—INDONESIANS—NEGRITOES—PAPUANS.

To some extent the distribution of **the** human races throughout
the archipelago conforms **to** that of the lower organisms. Thus the
light types, of undoubted Asiatic origin, have their home in the
Asiatic or western, the dark in the Australian or eastern, division.
But the former, being more intelligent and enterprising, have somewhat encroached upon the domain of the latter. Hence the dividing
line between the two has been shifted considerably to the east, and
is drawn by Wallace in such a way as to transfer Lombok, Sumbawa,
Celebes, parts of Buru and Jilolo, with Tidore and Ternate, from the

eastern to the western division. But all along the frontiers of both
worlds there are blendings, overlappings, and intermixtures of all
sorts, while in the Philippines, in other respects mainly Asiatic, the
aboriginal element was clearly not light but dark. It is obvious that
man appeared much too late on the scene to be affected by the
original distribution of land and water ; as, for instance, in miocene
times. Hence the Oceanic division cannot here be taken into
account, and the presence of a dark people so far west as the
Andaman Islands, and till quite recently also in Java, might lead
us to suppose that the whole area was originally the exclusive
domain of this race. On this supposition the light-coloured people
would have to be regarded as everywhere intruders from the Asiatic
mainland, a conclusion which seems also to be justified on broader
anthropological considerations. In general it may be assumed that
the dark is the aboriginal, the light the intruding element throughout the whole of the Oceanic world, and consequently also in the
Eastern Archipelago.

It is commonly supposed that this region is at present occupied
by one light and one dark race only—the MALAYS in the west and
the PAPUANS in the east. But more careful observation has recently
shown that these are only the predominant races, and that beneath
them are two others, also respectively light and dark—the INDONE-
SIANS in the west, and the NEGRITOS, now restricted to the Philippines, and perhaps to some parts of New Guinea.

The Malays.—The affinities, general characteristics, and range
of the Malay peoples have been somewhat fully discussed in the first
part of this work, and here it will be sufficient to determine their
position in the Eastern Archipelago. The *Orang Maláyu*, or typical
Malays, who speak the standard Malay language, and who everywhere
recognise themselves as belonging to a distinct nationality, are
centred chiefly in the southern parts of Sumatra. Here alone they
form large and compact communities, such as those of Menangkábau
and Palembang ; here they first rose from the condition of rude and
savage tribes, developing a national culture under Hindu and more
recently under Mohammedan influences ; here, therefore, is the
true home of the Oceanic as opposed to the continental Malays, and
from this region they spread with the growth of trade and navigation
to various other parts of the insular world, which from them often
takes the name of the Malay Archipelago. Beyond South Sumatra
they are at present found settled chiefly round the coast of Borneo ;
in Tidore, Ternati, and opposite coast of Jilolo ; in Batavia, Singa-

pore, and all the large seaports of the archipelago ; lastly, at a few trading stations in Western New Guinea.

The so-called "High" or Standard Malay language has also been the chief medium of trade and general intercourse throughout the Eastern Archipelago, at least for the last four hundred years. This was due, not to any superiority of the Orang Malâyu over other members of the Malay family, but partly to the rapid spread in recent times of the Mohammedan religion through Malay traders and missionaries, partly to the softness and simplicity of the Malay language itself. The adoption of the general Arabic character, however otherwise unsuitable, was also a point in its favour, and it thus ultimately superseded the Javanese, Macassar, Bugis, and all other claimants for supremacy in the archipelago.

It is noteworthy that, with the exception of the Malays proper, all the cultured Malayan peoples, such as the Rejangs of Sumatra, the Javanese, the Bugis of Celébes, and even the Tagalas of the Philippine Islands, make use of peculiar writing systems, which are certainly antecedent to the introduction of the Arabic letters by the Mohammedans. While differing greatly in appearance, the alphabets resemble each other in their general characteristics, all running from left to right in horizontal lines, and being somewhat of a syllabic type. This points at a common origin of these orthographic systems, which have in fact been traced to an Indian source. The prototype are probably the Buddhist letters as seen especially on the rock inscriptions of King Asoka, dating from about the third century B.C.

To the same fundamental Malay stock belong several other groups, which have had an independent historic evolution, which speak languages more or less intimately connected with the common Malay speech, and which in their physical appearance still betray their common descent from the Mongoloid peoples of Southern Asia. All stand thus related to each other much in the same way, for instance, as the various members of the Aryan family are related one to the other. They form the bulk of the population in North Sumatra, Java, the Lesser Sunda Isles as far as Sumbava, Celébes, the Philippine and Sulu Archipelagoes. Thus are constituted altogether five more or less distinct Malayan groups, which may be tabulated as under :

Orang Malâyu (Malays Proper) : Menangkâbau, Palembang, and Lampong in Sumatra ; Rhio-Lingga Isles ; Singapore, Bintang, Lingen, Banca, Biliton, Bornean Seaboard, Tidore Termate, and West Jilolo, scattered communities in all the trading places throughout the Archipelago.

Sumatran Group : Atjinese, Rejangs, Passumahs.

Javanese Group : Javanese proper, Sundanese (West Java), Madurese, Balinese, with the natives of Lombok, who call themselves Sasaks.

Celébes Group : Bugis, Mangkassaras, and others in Celébes, Muna, Bouton, Sumbawa, Sula (?).

Philippine Group: Tagalas, Bisayas, and others of the Philippines, the natives of Palawan and the Sulu Islands.

In all these the distinctly Malay physical type decidedly predominates. They are not, however, to be regarded as subordinate members of the Malays proper, but rather as independent branches of the common Malay stock. The Javanese group especially boasts a far older and far higher civilisation even than that of the Menangkábau Malays. Although now mostly Mahommedans, they had already adopted some form of Hinduism, probably three or four centuries before the new era, and under Indian influences had developed a very advanced state of culture nearly two thousand years ago, that is, at a time when the oldest of the Orang Maláya were still little removed from the savage state. Under a completely organised although despotic government, the arts of peace and war were brought to considerable perfection, and the natives of Java became famous throughout the East as accomplished musicians and workers in gold, iron, and copper, none of which metals were found in the island itself. They possessed a regular calendar with astronomical eras, and a metrical literature, in which, however, history was inextricably blended with romance. Bronze and stone inscriptions in the Kawi, or old Javanese language, still survive from the 11th or 12th century, and to the same dates may be referred the vast ruins of Brambanan and the stupendous temple of Boro-bodor in the centre of the island. There are no statues of Hindu divinities in this temple, but many are found in its immediate vicinity, and from the various archaeological objects collected in this district, and illustrated by A. B. Meyer of Dresden, it is evident that both the Buddhist and Brahminical forms of Hinduism were introduced at an early date. But all came to an end by the overthrow of the chief Hindu power in 1478, after which event Islâm rapidly spread over the whole of Java and Madura. Brahminism, however, still holds its ground in Bali and Lomboh, the last strongholds of Hinduism in the Eastern Archipelago.

From the Malayan groups must carefully be distinguished—

The Indonesians, who, although usually grouped with the Malay branch of the yellow Mongolic division of mankind, present rather the fair or light brown complexion and regular features characteristic of the Caucasic races. Such are the Battak in North Sumatra; the Kubus and Passumahs in Central and South Sumatra; the Mentawey Islanders, west coast of Sumatra; the Búlúdúpies of North Borneo; probably most of the indigenous inhabitants of Celébes; the Galelas of North Jilolo; many of the natives of Buru, Ceram, Savu, and Rotti; some of the Philippine Islanders, and the red-haired community recently met by H. O. Forbes in East Timor. They are everywhere found in the more inaccessible districts, and occupy a uniformly lower state of culture than the Malays, whom they appear to have preceded in the archipelago. Hence the term " Pre-Malay " applied to them by Dr. Hamy, although " Indonesian," originally suggested in a somewhat different sense by Logan, seems to be a more suitable designation. It serves to connect them with the brown *Polynesians* of the Eastern Pacific (Samoa, Tahiti, Hawaii, Tonga, Maori, &c.), who may be regarded as their descendants.

The relations of these two now widely-severed branches of the light-coloured Oceanic peoples become more and more evident according as more accurate knowledge is accumulated regarding them. Typical Indonesians are the Mentawey islanders, of whom Von Rosenberg remarks, that "as regards physical appearance, speech, customs, and usages they stand almost quite apart. They bear such a decided stamp of a Polynesian tribe, that one feels far more inclined to compare them with the natives of the South Sea Islands."—*Malay Archipelago*, l. p. 189. These and the other Indonesians are described as of a somewhat light ruddy-brown and even fair complexion, with long wavy or curly hair, black or inclining to a brown shade, beard often fairly developed, well-modelled torso, large muscular frame, rather above the middle size, dolichocephalic, or long head, full, open, and horizontal eyes, high forehead, straight nose, and regular, oval features. This description at once separates them from the low-sized, round-headed, oblique-eyed, lank-haired, short-nosed, yellow Mongoloid Malays, and seems to affiliate them, on the one hand, to the large, brown, eastern Polynesians, on the other, to the swarthy or fair and regular-featured western Caucasic peoples. To account for these resemblances it is only necessary to assume a remote migration of the Caucasic race to south-eastern Asia, of which evidences are not lacking in Camboja and else-where, and a further onward movement, first south to the archipelago, and thence east to the Pacific. The problem is fully discussed in A. H. Keane's *Relations of the Indo-Chinese and Inter-Oceanic Races and Languages*.

Negritoes.—The Negritoes, that is, in Spanish, "Little Negroes," are now confined mainly to the Philippines, and even here survive only in the five large islands of Luzon, Mindoro, Panay, Negros, and Mindano, numbering altogether probably not more than 20,000 souls. They are collectively known by the name of *Aëta*, or *Ita*, which in Tagala means "black," answering to the Malay *Hétam*. Their affinities are with the Samangs of the Malay Peninsula, the Andamanese Islanders, the Karons of New Guinea, and the Badui and Kalangs of Java, with whom they have in common a dwartish stature, seldom exceeding four feet six or seven inches, a brachy-cephalic or round skull, very short frizzly or woolly hair, said to grow in separate tufts (?), short nose, thickish lips, and generally a somewhat infantile Negroid expression. Further exploration may reveal the existence of true Negrito tribes in Celébes, Jilolo, Timor, and Borneo, although it now appears that none survive in Formosa, where their presence had long been suspected. De Quatrefages finds traces of a Negrito element in Southern India, on the slopes of the Himálayas, and as far west as Sistan on the Perso-Afghán frontier.

But in any case their survival at such widely separated points as the
Andamans, the Philippines, and New Guinea seems to justify the
commonly-received opinion that they are the scattered fragments of
an aboriginal dwarfish Negro race, formerly diffused over the Eastern
Archipelago and adjacent Asiatic seaboard. Before the total subsidence of the Lemurian continent, their range may have even
extended to Africa, where dwarfish Negroid peoples, such as the
Akkas, Obongos, and Bushmen also still represent the *disjecta
membra* of a primitive black pigmy element at one time spread over
a great part of the African mainland.

Characteristic of the Aëtas, as of all Negrito peoples, is an extremely
low stage of culture, which has never advanced beyond the hunting and
fishing state. They have no fixed abodes, or any dwellings beyond frail
structures of branches and brushwood ; their weapons are the bow and
poisoned arrow ; their food the products of the chase, roots, berries, and
vermin ; their costume necklets and armlets of beads and shells. Where
unaffected by Malay influences, their speech appears to be extremely rude
and undeveloped, broken into as many mutually unintelligible dialects as
there are tribes, and incapable of expressing any abstract ideas. But the
only Negrito language of which we have any adequate knowledge is the
Andamanese, which has been carefully studied by Mr. Man. Religious
notions are restricted to a dread of the surrounding spirits, which are
endowed with human faculties, though more powerful than ordinary
mortals. They lurk in the recesses of the hills, and flit about in the gloomy
forests, shaking the ground when angry, causing volcanic outbursts, and
bringing down the lightning from heaven. Of an after life there is no
thought, of the past no knowledge, all care being absorbed in the immediate present.

The Papuans.—The parallelism above suggested between the
African and Oceanic Negritoes applies with even greater force to the
African and Oceanic Negroes. The latter, familiarly known as
Papuans, from the Malay *papúwah* = frizzled, in reference to their
characteristic " mop-heads," are essentially a negro race, whose diffusion eastwards to the Pacific can also be best explained by the theory
of a Lemurian continent, or at least a chain of Lemurian islands
stretching across the Indian Ocean down to late tertiary times. The
disappearance of these lands, except at the two extremities, Madagascar
and Celébes, necessarily broke up the Negro family into two great
sections, and the separation took place at a sufficiently remote epoch
to account for the comparatively slight subsequent divergence of the
Western and Eastern types. This is perceptible chiefly in the nose
and mouth, which in the African have mostly retained the primitive
negro characteristics, but in the Papuan have become somewhat more
shapely and more conformable to a higher standard of physical

beauty. The Papuan nose is long, regular, arched or aquiline rather than concave, and tipped downward at the base rather than upturned. The nostrils also are narrower, the lips thinner and less protruding than those of the African Negro. In most other respects the types are similar, the Papuan having, like his Western congener, a long head, woolly hair, medium stature, or rather above the average, considerable muscular vigour, a light, cheerful disposition, and also unfortunately a decided taste for human flesh.

The present Papuan domain stretches across sixty degrees of the meridian (120°—180° E), from the island of Flores (Sunda group) to the Fijian Archipelago, and lies mostly between the equator and the Tropic of Capricorn. It thus comprises most of the islands east of Celebes, New Guinea, with all the adjacent groups (Key, Aru, Waigeon), the Louisiade Isles, New Britain, New Zealand, and the whole of Melanesia (Solomon, New Hebrides, New Caledonia, Loyalty, and Fiji Islands). In the Eastern Archipelago the dark populations between Flores and New Guinea are seldom of a pure Papuan type, almost everywhere betraying evidences of intermixture with the surrounding Malayan and Indonesian peoples. Hence they are called "Negro-Malays" by Crawfurd, who, however, unnecessarily regards them, not as the outcome of a fusion of those two elements, but as a separate race distinct from both. To them many writers apply the term "Alfuro," which is written in a variety of ways (Arfuro, Arafura, whence the Arafura Sea, &c.), but which has no real ethnological significance at all. In the mouth of a Malay "Alfuro" means simply wild, uncivilised, pagan, hence is indiscriminately applied to all unsettled, non-Mohammedan tribes at a lower stage of culture than the ordinary Malayan standard irrespective altogether of racial differences. The Galelas, for instance, of Jilolo, are "Alfuros," although, so far from being dark, they are a distinctly fair people of almost Caucasic type.

In the Papuan islands are current a very large number of languages, many of which also afford clear proof of Malayan influences. The numerals and words connected with trade, the arts, and industries are mostly of Malay origin. But the substratum is certainly distinct, as shown in the harsher phonetic system, the totally different structure, and the large number of independent terms expressing simple primitive ideas. The Malayo-Polynesian tongues certainly stretch from Madagascar across the two oceans eastward to Easter Island, and are spoken not only by most of the Indonesians in the Eastern Archipelago, but also by nearly all the Melanesians, or Papuans, of the Pacific. But it is not to be supposed, with Mr. Codrington, that, excluding Australia and the Negritoes, there are no other stock languages in this vast watery domain. The exploration of New Guinea and Borneo, scarcely yet seriously begun, will probably bring to light many fundamental forms of speech, and enough is known of the idioms current amongst the Papuan natives of Timor, Aru, Mysol, Nufor (Geelwink Bay) to show that several languages radically distinct from Malayo-Polynesian still survive in the Eastern Archipelago.

The Papuan populations have been carefully studied in recent times, especially by Wallace, A. B. Meyer, H. O. Forbes, and Miklukho-Maclay. From the varying and occasionally even contradictory statements of these and other observers it is evident that, with a certain general uniformity of physical type and mental qualities, there prevails a considerable diversity in the appearance, social usages, and general culture of the various branches of this race. Such discrepancies are to be attributed partly to the wide range occupied by them, but much more to the influence of the Malays, Polynesians, and other peoples with whom they have been in contact from the remotest times, especially on the smaller islands and around the sea-board of New Guinea. Thus it happens that while some are described as fairly intelligent, skilled husbandmen, endowed with some artistic taste shown especially in their curious wood-carvings, and altogether enjoying a considerable degree of culture, others appear to be very little or not at all removed from the purely savage state, land and sea nomads living entirely on the products of the chase, or on captives taken in the tribal wars, without fixed habitations, and ignorant of the most rudimentary arts. Miklukho-Maclay resided some time a few years ago amongst some communities on the north-west coast of New Guinea, who had no knowledge of the metals, all their implements being made of stone, bones, and wood. They did not even know how to make fire, so that when extinguished in a hut it had to be brought from another, or from a neighbouring village if extinguished in all the huts of the tribe at once. Their grandfathers told them of a time when they had no fire, and ate their food quite raw. They do not bury their dead, but place them in a sitting position covered with leaves of the cocoa-palm, the wife keeping a fire close to the **corpse** for two **or** three weeks till it is quite dried.

But apart from extremes, Wallace's classical description of **the** average Papuan may be accepted as fairly accurate. "The moral characteristics of the Papuan appear to separate him as distinctly from the Malay as do his form and features. He is impulsive and demonstrative in speech and action. His emotions and passions express themselves in shouts and laughter, in yells and frantic leapings. Women and children take their share in every discussion, and seem little alarmed at the sight of strangers and Europeans. Of the intellect of this race it is very difficult to judge, but I am inclined to rate it somewhat higher than that of the Malays, notwithstanding the fact that the Papuans have never yet made any advance towards civilisation. The Papuan has much more vital energy, which would certainly greatly assist his intellectual development. Papuan **slaves** show no inferiority of intellect compared with Malays, but rather **the** contrary, and in the Moluccas they are often promoted to places of considerable trust. The Papuan has greater feeling for art than the Malay. He decorates his canoe, his house, and almost every domestic utensil with elaborate carvings, a habit which is rarely found among tribes of the Malay race."—*Malay Archipelago*, ch. xl. Forbes also speaks of the high artistic ability of the Timor-laut Papuans, "very deft-fingered and clever carvers of wood and ivory. The figure-heads of their outrigger praus, dug out **of** single trees, especially attract attention by the excellence of the workmanship, carefully and patiently executed, and the elegance of their furnishings; while the whole length of their houses are also most elaborately carved with intricate patterns and representations of crocodiles and other animals. Their appreciation of beauty is a characteristic of them, which, absolutely wanting in the Malay people, I was surprised to find among a less advanced race."—*A Naturalist's Wanderings*, p. 317.

Of foreign nations by far the most numerous and socially influential are the CHINESE, who are **numerous especially** in Sumatra, Java,

Borneo, and the Philippines. They find employment as miners and
cultivators in Borneo, as petty dealers and labourers in all the large
coast towns, and as traders and seafarers almost everywhere. But
although a large share of the general commercial movement is in
their hands, and although their relations with the archipelago are of
long standing, they appear to have formed very few settled communities of a permanent character beyond Singapore and the Malay
Peninsula. Even where alliances of a more or less temporary
nature are formed with the native women, their chief ambition is
to make enough money to retire and spend their remaining years
amongst their friends at home. Even when these hopes are thwarted
by the incurable national vice of gambling, they still endeavour to
leave sufficient to have their bodies brought back for burial in their
native land. Hence the fears at one time entertained that the archipelago might become an Oceanic China are not likely to be realised.

Next in importance are the KLISES (Telingas) from the Madras Presi-
dency, whose position in the archipelago is somewhat analogous but less
commanding than that of the Banians of Bombay in Zanzibar and the other
trading places round the shores of the Indian Ocean. Some Arab communities are also found at various points, as in the Sulu Islands, where
M. Montano recently met a small group of nearly pure Arabs, but so
long settled in the country that they had lost all memory of their distant
homes.—*La Nature*, April, 1880. The Europeans, political masters of the
whole region, are nowhere numerous, and nowhere form permanent settlements in these tropical lands. They belong almost exclusively to the
official and military classes, and, like the Chinese, hasten to quit the
country as soon as they become entitled to retiring pensions.

CHAPTER IV.

GEOGRAPHICAL AND POLITICAL DIVISIONS—DUTCH, SPANISH, ENGLISH,
GERMAN, AND PORTUGUESE TERRITORIES—ASIATIC, OCEANIC, AND
AUSTRALIAN NATURAL DIVISIONS.

POLITICALLY the whole of the Eastern Archipelago is distributed
amongst the five European States of Holland, Spain, Great Britain,
Germany, and Portugal. Some districts, such as parts of North
Sumatra, Borneo, Celébes, and New Guinea, are no doubt practically
beyond the control of any foreign power; but their autonomy is
scarcely anywhere recognised, so that for administrative purposes
the archipelago must be regarded as a dependency of Europe. The

distribution, however, is effected in an extremely irregular manner, for while Portugal claims only a section of the comparatively unimportant island of Timor, the preponderance of Holland is so great, that the expression " Dutch East Indies " might almost be applied to the whole region in the same way that " British India " is understood to comprise the whole of the Indian Peninsula. On the other hand, the political arrangement so completely over-rides all natural or physical considerations, that it cannot be attended to in any intelligible description of this insular world. To do so, Borneo and Timor, for instance, would have to be treated under two, and New Guinea under no less than three separate heads. Hence in the subjoined detailed account of the archipelago the three geographical divisions—*Asiatic, Australian,* and *Oceanic,* as explained in chapter I., will be adhered to, political interests being consulted by the full tabulated statements of the several European possessions, which will be found in the statistical tables. In supplement to these tables it may **here be** stated, in a general way, that **to** Spain belong exclusively **and** solely the *Philippine* and *Sulu* groups ; to England the northern section of *Borneo,* the islet of *Labuan* off the North Bornean Coast, *Keeling Islands,* and the south-eastern section of *New Guinea* east of the 141° E. longitude ; to Germany the north-eastern section of *New Guinea* east of the same meridian ; to Portugal the eastern and smaller section of *Timor ;* to Holland all the rest. The relative extent **and** population of these possessions are as under :—

	Area in Sq. Miles.	Population.
Dutch . . .	740,000	27,429,000
Spanish . .	115,000	5,636,000
British . .	113,000	291,000
German . .	70,000	109,000
Portuguese . .	6,000	300,000
Total	1,044,000	Total 33,765,000

1. ASIATIC DIVISION.

The Large Sunda Group, with Bali and islands adjacent to Sumatra —The Philippine and Sulu Archipelagoes.

THIS division, comprising considerably more than half of the Eastern Archipelago, or about 570,000 square miles, lies mainly in shallow waters, seldom exceeding 100 fathoms, except towards the north-east. Here a deep trough in the China Sea, combined with other indications, shows that the Philippines were detached from the Asiatic mainland at a much more remote geological epoch than the large Sunda group, that is, the great islands of Borneo, Sumatra, and Java. The shores of these insular masses facing the Malay Peninsula and Indo-China are elsewhere washed by shallow inland seas, which were probably dry land so recently as early pleistocene times. But on the opposite sides great depths are soon reached, so that the original Asiatic coastline is indicated by the chain of the Nias and Mentawey islands running parallel with the west coast of Sumatra, and thence by a line drawn within twenty miles of the South Javanese and East Bornean seaboards. All the lands enclosed by this curved line rest upon a submarine plateau with a mean depth of little over thirty fathoms, so that a slight upheaval of about 200 feet would suffice again to connect them with the mainland, to which their geological and biological features also show that they originally belonged.

BORNEO.

Borneo, the most central and next to New Guinea the largest island in the Eastern Archipelago, has no general native designation, although by the Malays sometimes called *Tanah* or *Pulau Kilémantan*, "Land or Island of the Mango." The name by which it has been known to Europeans since its discovery is merely a corrupt form of *Brûnei* (*Brûnai, Brûni*), capital of a still existing Malay State on the north-west coast, which was the first place visited by the Magellan expedition in 1521. It is nearly bisected by the equator, lying between 7° N. and 4° S. lat., 109°—119° E. long., with the China Sea to the north and west, Macassar Strait to the east, and the Java Sea on the south. Its greatest length, 690 miles, is almost exactly indicated by the 115th meridian running from Point Sampan-Mangio at Marûdar Bay southwards to Tanjong Selátan near the Banjer River in the Banjer-Massin Residency. Its greatest breadth, 605 miles, lies in lat. 1° N. between the mouth of the Sambas river below Sarâwak and Point Kanyungan in Macassar Strait. It presents a somewhat massive quadrangular form unlike that of any other large island in the world, with a total area of 263,000 square miles as measured on Brinkman's large map of 1879, but by other authorities estimated as high as 290,000 and even 300,000 square miles. The estimates of the population, based largely on mere conjecture, present still greater discrepancies, varying from about 1,750,000 to 2,500,000 and upwards.

With a somewhat irregular coast-line of over 2000 miles, but less indented than that of most other large islands in the archipelago, Borneo offers few spacious bays or deep-water havens, except in the north, where the seaboard is higher and more abrupt. Elsewhere the shores are mostly fringed by a broad margin of low-lying and marshy lands from 30 to 50 miles wide, mainly of recent alluvial formation. New land, as in Landak on the west side, is known to have been gained from the sea during the last four hundred years, and the coast-line appears from other signs to be extending seawards.

The generally uniform outlines are relieved chiefly by Datu Bight on the west; Brunei and Marúdu Bays with Kudat harbour on the north; Paitan, Labuk, and Darvel Bays with Sandákan harbour in the extreme north-west; Adang, Pamukan, and Klumpang Bays on the west, and Sampit Bay on the south coast. The most conspicuous headlands are Capes Datu, Sirak, and Baram on the east side; Sampan-Mangio and Unsang on the north; Kanyungan on the east; Sungei-Bharn, Malang-Layer, Flat Point, and Samba on the south.

There are few important islands on the Bornean seaboard, those which formerly existed having apparently been joined to the mainland by the process of upheaval, or creation of new alluvial land now going on. The largest are Pulau Laut, close to the south-east coast; Caramata, which gives its name to the channel between the south-west coast and Billiton; the Tambilan and Natuna groups far to seaward of the west coast; Banguey (Banggi) and Balambangan, ten miles from the northern extremity of Borneo. Some historic interest attaches to Balambangan, where the East India Company made its earliest settlement in these waters in 1763, over twenty years before the occupation of Pinang. Since the recent creation of British North Borneo, Balambangan has again become British territory.

Of the interior of Borneo a large part still remains to be explored, so that of its general configuration little is known beyond the more salient features. The mountain system seems to be so disposed that, were the land submerged a few hundred feet, it would present somewhat the same curious outlines as the more westerly islands of Celébes and Jilolo. From a central nucleus lying much nearer the west than the east coast there radiate north, east, and south four main ranges, enclosing three broad plains, or slightly elevated tablelands, corresponding to the three great marine inlets on the east side of Celébes.

Under the various names of Kelingkang (Bayang-Miut), Madi, and Anga-Anga, the largest and loftiest of these ranges traverses

Borneo from Cape Datu on the west coast to Mount Kina Balu
(13,698 feet ?) in the extreme north-east, which is usually supposed to
be the culminating point of the whole island. But Carl Bock heard
from the Kûtei Dyaks of a great central chain called the Tibang,
which is the common source of the Kapuas, Mahakkam, and other
great rivers, and which may contain peaks higher even than Kina
Balu. This explorer also heard of a Mount Tepu-Purau, "so high
that it is said to be within a trifle of reaching heaven."— *Head-
Hunters of Borneo*, p. 176.

From the central nucleus diverge other ranges, such as Kaminting
southwards, Sakûru and Merâtu towards the south-east, enclosing
wide lowland plains and in some parts rising to a height of 6000
feet. The prevailing rocks are limestone, slate, sandstone conglomerates, and on the mountain tops syenitic granite. Of active volcanoes there is no trace, although the southern and western coasts are
little over 200 miles distant from the great volcanic belt passing
through Sumatra and Java. Many of the Bornean peaks, however,
bear distinct evidence of former igneous activity, and some of the
cones appear to have been regular craters in remote geological times.
They were probably active during the paleozoic and early mesozoic
periods, to which Mr. Tenison-Woods refers the vast coal measures
forming a leading geological feature of the island.

"There are few countries in the world," says this naturalist, "except,
perhaps, Eastern Australia, where coal is so extensively developed as in
Borneo. Thick seams crop out in innumerable places on the coast and on
the banks of the rivers. In some of the streams of North Borneo I have
seen water-worn and rounded fragments of coal forming the entire shingle
of the channel. In some places, again, there are outcrops with seams of
good coal twenty-six feet thick. The coal-formation is the one prevailing rock of the coast. It forms the principal outcrop at Sarâwak ; at
Labuan, also, no other rock can be seen. All the grand scenery of the
entrance to the port of Gaya is made up of escarpment of coal rocks. At
Kindat it is the same, and so I might go on with a long list of coal-bear-
ing localities."—*Nature*, April 23, 1885. In many places the coal is of
excellent quality, quite bituminous and not brittle, and the mines now
being worked by two Scotchmen in the Brunei district are of great value.
　Other mineral products are gold, occurring especially in the west (Sarâwak, Sambas, and Pontiânak), antimony found in great abundance in
Sarâwak, mercury and iron, the latter in Kûtei and elsewhere ; lastly,
diamonds in Sarâwak, Pontiânak, Landak, and Martapura. Borneo is the
only island of the archipelago which has yielded these crystals, the largest
in the world being one of 367 carats, if it be a genuine diamond, belonging
to the Rajah of Matan. The doubt which rests on this famous stone is due
to the fact that its owner will not have it cut, and refuses to submit it to
the examination of experts. It was found about the year 1787 in the

Landak mines near the west coast, which are amongst the oldest and most productive in the world.

Rivers.—Borneo is abundantly supplied with rivers, which may be disposed in five main fluvial basins. Of these the least extensive comprises the north-western slope of the Kelingkang range, draining the Brunei and Saráwak districts through the Tewáran and Tampásuk rivers, having their sources in Kina Balu ; the Limbang (Brunei), Baram, Bintûlu, Rejang (navigable for 140 miles), Seríbas, Bátang-Lupar, and Saráwak. But by far the largest rivers are those of the south-western basin, where the Kapûas, rising in 114° E. long., reaches the coast between Mempawa and Sukadâna, and the Baríto (Banjer-Masin), the great fluvial artery of the island, flows for hundreds of miles nearly due south to the Java Sea in 114° E. long. over against Madura. Both of these streams appear to have a common source near the unexplored Lake Kûtei-Lama.

Next in importance is the Kûtei, or Mahakkam, which rises in Mount Lasan-Tula, and flows with a rapid course mainly eastwards to Macassar Strait, which it enters through several mouths. Its delta, projecting considerably seawards, and embracing 50 miles of coast, presents great difficulties to navigation, although the main channel is accessible as far as Mûra Pahou (116° E. long.) for steamers drawing 10 feet of water. According to Carl Bock, the Mahakkam has a total length of not less than 600 miles. North of the Kûtei basin is that of the Bulangan, which has also an easterly course, with a delta in the Celébes Sea at the entrance of Macassar Strait. Besides these there are hundreds of smaller streams reaching the coast in independent channels, but mostly inaccessible to vessels of heavy draft.

Lakes.—No large lakes, strictly so-called, are known to exist in Borneo, those spoken of by travellers being rather temporary lagoons, or expansions of the rivers in the low-lying or level plains. Even the existence of the much-disputed Lake Balu, south-east of Kina Balu, has recently been disproved. Mention, however, is made of a Lake Danau-Maláyu, in 1° 5′ N. lat., 114° 20′ E. long., said to be over 20 miles long, and 10 or 12 broad, with a depth of 18 feet.

The formation of the Bornean lakes, or "danaus," as they are called by the natives, is thus described by Dr. Schwaner :

"By danau is to be understood an inland sheet of water in the deepest part of a marshy district in the immediate vicinity of rivers. Their mode of formation often resembles that of the canals, such as those of the Martapûra, and other Bornean streams, which are used for shortening the water route, and sometimes dug by hand, but mostly formed by the rush

K

of water during the floods. Channels thus formed eventually expand to a
danau, the water at every flood flowing in and enlarging its margins. The
lakes have no determined shores, the ground sinking imperceptibly to its
greatest depth, while the constant shifting of the land surface causes a
corresponding expansion or contraction of the lacustrine area" (Borneo).
Some of the Mahakkam lagoons are over 15 miles in length, and figure on
our maps as true lakes of old geological formation.

Climate.—The rainfall is remarkably heavy in most parts of
Borneo, but especially in the north-west, where it averages 180 inches
at Saráwak. The climate, however, owing to the influence of the
sea breezes, is milder and healthier than in most other islands of the
archipelago, although Bock found it very insalubrious in some parts
of the interior, where hot land-winds probably prevail. The west
coast has no really dry season, being refreshed by heavy and continuous rain throughout the year, and especially from December to
March. The mean reading of the thermometer is 82° F. at Pontiának,
where it never exceeds 92°. But in North Borneo, Guillemard
recorded 95°, and found the heat in the Sigalind valley "almost as
trying as that of New Guinea or West Africa." (*Cruise of the Marchesa,*
ii. p. 95.)

Flora and Fauna.—Except in some alluvial districts, the soil
of Borneo appears to be less fertile than that of the volcanic islands.
Much of the surface is still covered with a primeval forest growth,
including not only the gigantic timber, such as ebony, ironwood,
sandalwood, &c., which the poorest soil will produce in equatorial
regions, but many of the most useful tropical species, such as benzoin,
camphor, gutta, the sago-palm, and the rattan. The latter thrives
especially in the south-west, the rattan of Banger-Masin having a
higher value than that of any other country in Malaya. The chief
cultivated plants are rice, maize, cotton, opium, pepper, yams, and
indigo. The mangosteen, durian, and many other delicious fruits
abound.

The most remarkable feature in the zoology of Borneo is the absence or
rarity of many large animals found in the adjacent islands. Thus the
tiger and leopard of Java and Sumatra are unknown, their place being
supplied by a smaller species, the *Felis Macrocelis.* Scarcely less remarkable is the absence both of the elephant and rhinoceros from by far the
greater part of the country. Of large animals, the most interesting are
the orang-outan (next to the gorilla the largest anthropoid ape), the wild
cattle, and the proboscis monkey. Deer, wild swine, and squirrels are
everywhere abundant. Other characteristic mammals, not, however,
peculiar to Borneo, are the honey-bear (*Ursus Malayanus*), the binturong,
intermediate between the civet and bear, the tiger-cat (*Felis planiceps*),
the Kubury, or flying lemur (*Galeopithecus volans*), the curious silat, or
gobang (*Mydaus meliceps*), a kind of badger with a pig's snout, and the

scaly ant-eater (*Manis javanica*, or tangiling of the Malays), which rolls
up like a hedge-hog. In its birds Borneo agrees very closely with Sumatra,
the peacock being absent, while the argus and fire-backed pheasant abound.

Inhabitants.—Of the inhabitants of Borneo probably one half,
or about a million, belong to the aboriginal stock collectively known
as DYAKS, and usually regarded as a branch of the Malay race. But
many are of such fair complexion and regular features that they
should perhaps be grouped rather with the Indonesian family. They
are divided into a vast number of tribes, speaking more or less
distinct languages, mostly of the Malayo-Polynesian type; but they
differ greatly from the Malays in other respects, being much more
lively and active, and of a more trusting disposition, while many
"have, notwithstanding their high cheek-bones and broad noses, a
type of face which is quite in accordance with European ideas of
beauty."—C. Bock. The Dyaks of Long Wai, Long Wahou, and
many parts of Kûtei are above the average height, while the Kayans
of the interior, about the headwaters of the Rejang and Bintûlu and
elsewhere, present peculiarities distinguishing them both from the
Malays and ordinary Dyaks. They are supposed to be an intruding
race, which came originally from Celebes, and penetrated from the
east coast far into the interior. Although considered one of the most
advanced of uncivilised races, they are addicted to head-hunting,
which, however, is prevalent amongst most of the native tribes.
Apart from this propensity, the Dyaks are described as very honest,
respectful, and kind to their women, fond of their children, and of
temperate habits. But some of the tribes in Kûtei and elsewhere
are decided cannibals, and at Muera Pahau Carl Bock made the
acquaintance of Sibau Mobang, chief of a man-eating community,
who had recently slaughtered and devoured seventy victims. He
stated, however, that his people did not eat human meat every day,
that being a feast reserved for head-hunting expeditions. At other
times their food consisted of the flesh of various animals and birds,
rice and wild fruits, to which the ordinary Dyak adds fish and
curry.

Owing to its universality and intimate association with religious rites,
head-hunting is the most distinctive feature, as well as the plague-spot of
Dyak society. Births and "namings," marriages and burials, besides many
less important events, are all accompanied by marauding expeditions to
some neighbouring tribe for the purpose of securing a few heads to add
solemnity to the festivity. Hence, in the more inaccessible districts of
the interior, the practice has full sway, and is regarded as a chief cause of
the steady diminution of the population. No youth can be married, or
associate with the opposite sex, until he has taken part in one or more

head-hunting expeditions, and the more heads he can lay at the feet of his
beloved, the more he is admired by her and feared by his fellows. The
practice is not confined to Borneo, or even to the neighbouring islands,
but has flourished from remote times among many of the wild tribes in
Further India, and so far attests the continental origin of the "fair" races
in the archipelago.

Besides the Malays, who are restricted to a few centres near the coast,
several other peoples have settled in Borneo, which, from its central position, has naturally been made a resort for all the surrounding lands. That
the Javanese formerly made regular settlements in the south is shown by
monuments still existing as far north as the Kútei Valley. The Bugis of
Celébes have long maintained considerable settlements in the southern and
eastern districts. Further north are some communities from the Sulu
Archipelago, who formerly held a considerable tract of country about Cape
Unsang, and whose sultan till recently claimed jurisdiction over that part
of the island.

But the most important intruders are the Chinese, who are found
in every centre of population as traders, miners, mechanics, or
agriculturists. They are most numerous in the western districts,
where gold and diamonds are found, and there are said to be nearly
350,000 in the Dutch territories alone. The trade of Pontiânak,
Banjer-Masin, Saráwak, North Borneo, and Labuan is to a great
extent in their hands. But from old records of travel, the north-
eastern districts would seem at one time to have been even more
permanently occupied by the Chinese than at present.

Political Divisions.—Politically Borneo is distributed in very
unequal proportions amongst **the** Dutch, British, and **natives as**
under :

DUTCH POSSESSIONS IN BORNEO.

The Dutch claim sovereignty over the greater part of the island,
including the whole region south of a line drawn from Saráwak
north-eastwards to the source of the Sibuko river, about 4° N. lat.,
and thence eastwards to the coast a little above that parallel. Their
possessions are divided administratively into the three residentships,
of Pontiânak, Banjer-Masin, and the semi-independent State of
Kútei, with the usual system of residental government over the coast
districts of the west and south. But probably not more than one-
fifth of this region is under their direct control, a great part of **the**
interior being unoccupied and even unexplored.

Pontiânak.—In this residency are comprised the western districts of Sambas, Montrado, Sintang, and Pontiânak proper. The
town of *Pontiânak*, which lies on the west coast under the equator, is

the oldest trading settlement in Borneo ; but the district still remains under the nominal rule of a native sultan. Tin, gold, and diamonds are obtained from the numerous mines of this district worked by the Chinese. *Sintang* on the Upper Kapúas river is the seat of an assistant-resident, whose jurisdiction **extends** over the numerous Dyak tribes of the interior.

Banjer-Masin.—This is by far the largest residency, including the greater part of Borneo south of the equator, the **Kwân** country, and the sub-residencies of Amuntai and Martapúra, **with** a total population of about 600,000, mostly Dyaks. The capital, *Banjer-Masin*, lies some fifteen miles up the estuary of the Barito river, which is occupied by Dutch forts for 200 miles as far as Lútontúr, at the Teweh confluence near the equator. Thus the whole of the extensive Barito basin is securely held, although in many places the direct authority of the Dutch extends very little beyond the range of the strategical stations. In the interior, the most important trading place is *Bakompai*, one of these stations about sixty miles up the Barito at the confluence of the Nagara, its great affluent from the north-east. Here are collected most of the timber, gold-dust, gutta, resin, wax, edible birds'-nests, rattans, damar, and other local produce forwarded for export through Banjer-Masin. To the latter place the neighbouring district of Martapúra also sends its diamonds, the chief commodities received in exchange being European wares, such as coloured prints, cotton cloths, beads, and copper wire, besides tobacco, opium, salt, gambier, and Chinese earthenware. Although not so populous as Pontiának, Banjer-Masin is at present the largest, as well as one of the oldest trading ports in Borneo. On the south coast the chief seaport is *Tabunian*, **not** far from the capital.

Kûtei, or **Coti**, the eastern residency, comprises **the** extensive fluvial basin of the Kûtei (Mahakkam) river, with a **total area** of little less than 40,000 square miles. There is **an** assistant-resident at the Bugis settlement of *Tengârong*, thirty miles up the main stream, and the whole east coast from Sebambam **in** Tanah Bumbu **to** Kanyungan in 1° 3′ N. lat. is directly under the Dutch Government. Nevertheless, the native Malay sultan, Mohammed Solimân Kalifat **ul** Mumenin, who resides at Tengârong, still exercises the royal functions, and even possesses autocratic jurisdiction over his Mohammedan and Dyak subjects. While recognising the Dutch suzerainty, he keeps his own court, surrounds himself with various functionaries, levies taxes, and even inflicts capital punishment in a somewhat summary manner on his unruly subjects. Owing to personal antipathy to the Chinese, he encourages immigration of the enterprising Bugis from South Celèbes, who are slowly developing the resources of the country. Along the low and swampy east coast there are no natural havens, so that *Samerinda* and *Tengarong*, the only trading places of any importance in the Kûtei residency, are both situated somewhat inland above the Mahakkam delta.

British Settlements in Borneo.

Saráwak.—This territory, which now extends 400 miles east and west from Mount Mûlu (9000 ft.) to Mount Poi (6000) with a mean breadth inland of nearly 100 miles, has a total area of some 40,000 square miles, with a mixed population of 300,000 Dyaks, Dusuns, Malays, Chinese, and others. It lies on the north-west coast, and is watered by the Rejang, Baram, Batang-Lupar, and several other streams, some of which are navigable for a considerable distance inland.

The government of this territory, which has now lasted over forty years, and seems firmly established, was acquired from the sultan of Brûnei in 1841 by Sir James Brooke, better known as Rajah Brooke of Saráwak. In 1861 a second cession of territory was obtained including all the rivers and land from the Samalaran river to Kadûrong Point; and in 1882 a third cession of 100 miles of coast-lands, with all the riverain tracts between Kadûrong Point and the Baram, or rather three miles to the north-east of that river. The present rajah, H. H. Charles Johnson Brooke, who is a nephew of Sir James, succeeded in May 1868.

The success of this undertaking was shown during the Chinese insurrection in 1857, when the whole indigenous Dyak and Malay population rallied round the English Rajah, drove out the unruly miners, and triumphantly restored his power. By persevering in a liberal and enlightened policy, the rajah's government has brought peace and safety and comparative prosperity in the place of piracy, tribal warfare, and oppression.

Kuching, the chief town, is a thriving place of some 20,000 inhabitants, with the "Astana," or rajah's palace, and the bishop's house, the ordinary residence of the Diocesan of Singapore, Labuan, and Saráwak. It lies about twenty-three miles up the Saráwak river, which has a bar like other streams, but is accessible to this point for small steamers. The trade is chiefly with Singapore, to which it exports gold, silver, diamonds, antimony, quicksilver, coal, gutta, india-rubber, canes and rattans, camphor, wax, birds'-nests, sago, tapioca, pepper, gambier, and other local produce.

The government of Saráwak may be termed constitutional, resembling

in its main features that of a crown colony. The rajah is the absolute
head of the State, with full power of spontaneous and independent action,
which, however, he rarely exercises, being usually advised on local matters
by his Legislative Council of two Europeans and five Malay chiefs. A
general triennial assembly of the principal native and European represent-
atives of the several districts is held, and sometimes specially summoned
on urgent occasions. Any important change in the law or modification of
native custom would be referred to this General Council.

The government of the various districts, outposts, and riverain forts is
mainly entrusted to European Residents, aided by Assistant-Residents,
native, Eurasian, or Chinese clerks. The European staff now numbers
about twenty altogether. Although a mild system of slavery still exists,
the general tendency of the government has been to gradually reduce it
within the narrowest limits, with a view to its total abolition, which is now
imminent.

British North Borneo.—Another remarkable experiment in the
administration of uncivilised communities is exemplified in British
North Borneo, or the territory of Sabah, lately acquired by a chartered
trading company from the sultans of Brúnei and Sulu. Originally stretching from the Kumanis to the Sibuco river on the east
coast, with a seaboard of about 500 miles, it has been quite recently
extended towards Brúnei, so as to include the river Padas, with
an area of 26,000 square miles, and a mixed population of 200,000
Dusuns (half-caste Chinese and natives), Dyaks, Bajaus, Ilanos,
Búládúpies, Sulus, Malays, and Chinese. Occupying the north-
eastern corner of Borneo, between 4° and 7° N. lat., and 116°—119° E.
long., it is washed on three sides by the China and Celebes Seas, and
bounded southwards by Brúnei and Dutch territory; but not more
than forty or fifty square miles of this extensive region have yet been
settled, chiefly round Sandákan Bay on the east side. Here is the
town of *Sandákan (Elopúra)*, present seat of government, other
rising stations being *Kudat* on Marúdu Bay; *Abai* on Tampasuk
Bay; *Papar*, *Kimánis*, and *Gaya* on the west coast. The last-
mentioned gives its name to one of the finest natural harbours in the
archipelago, said to comprise ten square miles of good anchorage in
depths of from five to twenty fathoms. It is well sheltered from all
winds, might be easily defended from attack, and occupies a convenient commercial position near the trade route between Honkong
and Singapore, about midway between those great seaports. But at
present most of the export trade is centred in Sandákan Bay, which,
though much shallower than Gaya, is even better sheltered, being for
twenty miles almost completely landlocked.

North Borneo enjoys a much cooler climate than most places so
near the equator, the extreme summer heat rarely exceeding 85° F.

in the shade, while in the cool season the glass falls to 65°. The country is well watered by the Segama, Kimánis, Papar, Tampasuk, and many other streams, and its fertile soil yields good crops of rice, yams, sago, arrow-root, pepper, betel, and tobacco. Cacao, coffee, the cocoa, palmyra, and areca palms, are also cultivated, while the forests abound in ebony, camphor, bilian (ironwood), gutta-percha, india-rubber, rattans, and cinnamon. Although it is less rich in minerals than Saráwak, gold is reported to abound, especially in the Segama river basin, while coal appears to exist both on the east and west coasts. Traces of tin have also been discovered, and several extensive pearl fisheries are comprised within the company's boundaries. But at present the export trade is mainly restricted to camphor, ebony, rattans, sago, and edible birds'-nests. A powerful saw-mill is now at work at Elopúra, which has become the company's chief trading place. It stands on a headland commanding the approach to Sandákan Bay.

In connection with this promising enterprise will always be prominently associated the names of two persons—*Baron Overbeck*, to whom is due the first conception of the scheme, and *Mr. Alfred Dent*, who supplied the means of carrying it out. The Provisional Company was formed in 1877, and her Majesty's Charter was signed on November 1, 1881. Three principal establishments or residencies have already been organised, on the north-west coast, on the Papar and Tampasuk rivers, and at Sandákan Harbour on the east coast. The government is conducted by a governor, assisted by a council and by residents, the usual administration of a British Crown colony being adhered to as far as practicable. But although enjoying a charter, North Borneo has not yet been formally recognised as a British protectorate.

Labuan.—This island, which was purchased by the British government from the sultan of Brúnei in 1847, lies six miles off the north-west coast of Borneo in 5° 16′ N. lat., 115° 15′ E. long. It is twelve miles long and five or six wide at its broadest part, with an area of a little over thirty square miles. The settlement was formed mainly on account of the rich coal deposits, which have nevertheless proved the ruin of three companies, and are now little worked. The output fell from 6000 tons in 1876 to 800 in 1882 (Guillemard).

Labuan has a fine port, and has become a market for the produce of the neighbouring coasts of Borneo and the Sulu Archipelago, such as camphor, gutta, india-rubber, wax, pearls, tortoise-shell, birds'-nests, and túpang, forwarded mainly to Singapore. To the same place is sent the sago-flour prepared at three factories in the island from the imported raw material.

Labuan is administered by a Governor under the Colonial Office, who also acts as Consul-General for Borneo under the Foreign Office.

NATIVE TERRITORY IN BORNEO.

Brúnei, which gives its name to the island, is probably the oldest and the only surviving native State still enjoying absolute autonomy. It is a Malay principality on the north-west coast, bounded east and west by the British settlements of Sabah and Saráwak, and stretching from 100 to 150 miles inland, but with little authority beyond the coast. The natives are mostly Mohammedans, governed by a sultan, who is nominally absolute, and who until recent years ruled over the whole north-west coast from Saráwak to beyond Marúdu Bay. On the east side his territory was limited by the district round Cape Unsang, over which the sultan of the Sulu Archipelago claimed jurisdiction. But this potentate is himself now a vassal of Spain, while the Bornean portion of his territory is incorporated in British North Borneo. The last sultan of Brúnei died in 1884, in his hundredth year.

The largest and only important town in *Brúnei* is the capital of like name, which lies fourteen miles up the navigable river Limbang. Brúnei, described by Sir James Brooke as "a Venice of hovels," was seized by the British in 1846, but soon after restored upon the cession of Labuan. The population, more Malayan than that of other coast towns, was at that time stated to be 40,000, but had fallen in 1885 to some 25,000.

The State of Brúnei has become so disintegrated, that its ruler might probably be easily induced to surrender his royal prerogatives, and accept the position of a pensioner under the English Crown. The whole of North Borneo from Cape Datu to and beyond Cape Unsang might then be conveniently consolidated into a single British possession, capable of indefinite expansion under an enlightened administration. Its position on the great highway of trade between India and China could not fail to secure a permanent market for its abundant and varied natural resources.

HISTORICAL NOTES.

The growth of European influence in Borneo has been less steady and more intermittent than was the case in the Straits or in Java. The island was first visited by Lorenzo de Gomez in 1518, and by Pigafetta with the ships of Magellan's expedition in 1521. Both named it Brúnei from the sea-port on its north-west coast, where they happened to touch, and this term, written by the Malays *Bruni* or *Burni*, assumed the form of *Borneo* amongst Europeans, by whom it was commonly applied to the whole island in the 17th century. The alternative Malay expression, Tanah Kelamantan, or "Mango Land," may perhaps be current especially in the Dutch territory; but the native tribes have never had any general name for the whole island.

Commercial relations were soon formed with the natives by the Portuguese, at first in Brúnei itself, and then in the other maritime States. In 1573 the Spaniards, recently established in Manilla, endeavoured to open a connection with Brúnei, whose sultan had sought their aid, and was reinstated by them in 1580. But neither Spain nor Portugal ever exerted much real influence in the island, and the only missionary effort recorded in the sixteenth century ceased with the death of its promoter, the Theatine monk, Antonio Vintimigli.

Early in the seventeenth century the Dutch and English begin to appear on the scene, and in 1608 Samuel Blommaert was appointed Dutch resident in Landak and Sukadana. The English, who first visited Borneo about 1609, had a factory at Banjer-Masin in 1706. But from this they were soon expelled, apparently by the influence of the Dutch, who shortly after obtained a monopoly of the trade. The Dutch power became predominant round the west and south coast, when the rajah of Bantam had ceded his sovereign rights to their Company, and especially when the sultan of Banjer surrendered his territory about 1787.

The attention of the English was in the latter part of the eighteenth century turned towards North Borneo, then subject to the sultan of Sulu, from whom, in 1756, Alexander Dalrymple had obtained formal possession of Balambangan Island and all the north-eastern promontory. But the military post stationed here in 1763 was surprised and destroyed in 1775 by dains, or subordinate native chiefs dissatisfied with the cession of their territory. The Dutch also were overtaken by a series of misfortunes caused by their own mismanagement, and in 1809 all their settlements were abandoned by order of Marshal Daedels. The natives along the coast now resorted more and more to piracy, rendering legitimate trade so impossible that the settlement which the English East India Company had again made at Balambangan in 1804 was abruptly abandoned within a few weeks. But in 1811 an embassy from the sultan of Banjer-Masin to the British Government then established in Java, secured the appointment of a commander and resident. An expedition was at the same time sent against Sambas, and a post established at Pontianak.

On the restoration of the Dutch possessions in 1816 all these arrangements were cancelled, and until 1842 a free and undisputed field was left to the enterprise of the Dutch Government. About half the kingdom of Banjer-Masin was ceded by the sultan in 1823, followed by further concessions in 1825. On the east coast also the sultan of Kutei acknowledged for a time the Dutch authority, but soon retracted, and has ever since maintained a semi-independent regal state. About 1830 the Dutch supremacy was generally repudiated, their troubles in Java having diverted their attention from Borneo. On the opening of Singapore nearly all the Bugis trade, formerly centred in Amboyna, was diverted to the Straits, and direct relations established with Sarawak and Brunei. Then the necessity of suppressing piracy became so urgent that Mr. Brooke, aided by the British traders, at last succeeded in procuring the co-operation of English cruisers for that purpose. This led to political intervention, and in 1846-7 Labuan was ceded by the sultan of Brunei, who also agreed to make no cession of territory to any nation or individual without British consent. The Dutch, thus checked in the north, concentrated their attention on the development of their influence on the south and east coasts. In 1844 the sultan of Kutei acknowledged their protectorate, and the area of their administration has since then steadily increased round the southern seaboard. At present they have a nominal suzerainty over two-thirds of the island, although scarcely one-tenth appears to be under their direct control or administrative influence.

The establishment of an organised government by Sir James Brooke in Sarawak, followed by the recent occupation of North Borneo by a chartered English trading company, secures the exclusive predominance of British influence throughout all the districts not actually claimed by Holland. Henceforth these two States will be held jointly responsible for the material development of Borneo, and the intellectual progress of its indigenous populations.

SUMATRA.

Next in size **to** Borneo **of** the Great Sunda **group, Sumatra** stretches for 1070 miles north-west and south-east **between** 6° N.— 6° S. lat., and 95°—106° E. long. It is thus, **like Borneo,** nearly bisected by the equator, and with an average **breadth of** over 200 miles, has a total area of about 128,000 square miles, or 8000 more than that of Great Britain and Ireland together. Washed on the west by the Indian Ocean, it is separated on the north-east by Malacca Strait from the Malay Peninsula, and at its southern extremity by the Sunda Strait from Java.

Islands.—Sumatra is fringed on both sides by numerous islands and insular groups, all of which appear to have originally **formed** part of the mainland. But while on the west Si Malu (Hog), Nias, Batu, North and South Pora (Mentawey), North and South Pagey (Nassau), and Engano clearly indicate the primitive Sumatran coastline towards the Indian Ocean, on the east side Bintang, Lingga, Sinkep, Banca, and Biliton belong, on the contrary, to the Malay Peninsula, of which they form a geological extension southwards. But a slight upheaval of perhaps about 50 fathoms would suffice again to connect all these groups with Sumatra itself to the Asiatic Continent, of which they are merely detached fragments.

Bintang and Lingga, with all the circumjacent islands and islets, form collectively the Dutch Residency of Riou-Lingga. They are mostly fertile, and carry on an active trade in pepper and timber through the chief town Riou, on an islet off the south coast of Bintang, the largest of the group. Here also resides the sultan of Riou, now a dependent of the Dutch Government. But of all the Sumatran islands, the largest and most important are Banca (130 miles long), separated by the strait of like name from the south-east coast, and Biliton, of circular form, 40 miles across, separated from Banca by Caspar Strait. Both of these islands, which resemble each other in formation, appearance, and inhabitants, form Dutch Residencies, and are noted for their rich tin mines, which have been worked for the Dutch Government chiefly by Chinese miners since 1709, yielding about 10,000 tons of metal yearly. The great tin formation, which commences in Tenasserim, and extends almost uninterruptedly through the Malay Peninsula to Banca and Biliton, here comes to an end.

In the extreme south-west, considerably beyond the geographical or natural limits of the Eastern Archipelago, lies the small but interesting coralline group of the *Keeling* or *Cocos* Islands, now attached to the

Government of the Straits Settlements. When officially visited in August 1885 by Mr. E. W. Birch, this group, memorable in connection with Darwin's studies of atoll formations, was found to have already recovered from the effects of the terrific cyclone of 1875. All the islands (over 20 in number) are thickly planted with cocoa-nuts, and their Malay and Javanese inhabitants, 516 in 1885, do a brisk export trade in copra, cocoa-**nuts**, cocoa-nut oil, bêche-de-mer, and mendnku, a bark used for dyeing **purposes.**

Physical Features.—In its physical constitution Sumatra has much in common both with the adjacent mainland and with **Java,** while also presenting some features of an independent **character.** Along its west side it is traversed by the Barisan mountain system, parallel with but much more elevated than the main axis of the Peninsula. Lying in the line of the great volcanic belt, this range contains recent eruptive rocks like those of Java, and older plutonic and crystalline formations like those of the mainland. But Sumatra differs from both of these regions in the vast development of its plains, which mostly stretch from the western uplands right across the island to the eastern seaboard. In its general configuration it thus appears to be cast in broad and simple lines, mountainous and volcanic in the west, elsewhere level, and even low-lying and alluvial. "One may travel in some parts in a straight line westwards from the east coast for 150 or 200 miles without reaching an elevation of over 400 or 500 feet, while some 30 miles farther the Barisan peaks may ascend to over 10,000 feet."—H. O. Forbes, *op. cit.*

The culminating point appears to be Mount Kassoumba (15,000 feet ?), under the equator, other lofty summits being Indrapura (12,255) in Korinchi ; Lusé (11,000) in Achin ; Abong-Abong (10,000) : Telamon, **or** Ophir (9040) ; Salawanga (6,825). The volcanic cones are even **more** numerous than in Java, and the recent Dutch explorer, Verbeck, **states** that sixty-seven are known to exist, although two only (Merapi, **9640** feet, and Talang or Soelau, 8470) are still active. Other lofty cones are Kaba (9000 ?) and Dempo, in Palembang, both ascended in 1881 by H. O. Forbes, who determined the elevation of Dempo at 10,562 feet. This traveller also visited the Besagi and Tongamus volcanoes in the Lampongs district, both over 7000 feet. Thus the volcanic area is not confined to a small tract north and south of the equator, as is supposed, but extends from the extreme north to the extreme south of the island.

The great alluvial plain on the **east** side, lying not many feet above sea-level, and often extensively flooded, has a total length **of** 600 miles, with an average breadth of 60 to 110, and an area of **over** 40,000 square miles. But although this region is mostly **under** primeval forest, the eastern section of Sumatra is by no means of **such** uniform aspect as is generally supposed. The plains, table-

lands, and higher valleys, often of great extent, differ much in their
natural features, some being forest-clad and extremely fertile, others
arid and destitute of timber. Such especially is the already described
Pertibi plain in the Batta country, the aridity of this and other tracts
being no doubt due to the great elevation of the western highlands,
which are in some places lofty enough to intercept the rain-bearing
clouds rolling up from the Indian Ocean during the south-western
monsoons. As might be expected, the parched lands occur especially
in the north-east, where the mean elevation of the western range is
greatest, and where the uplands of the Malay Peninsula again form
a barrier against the moisture-laden clouds from the China Sea.
Hence, also, in the northern lowlands the intensely heated dry air
becomes rarified and replaced by cooler atmospheric currents, causing
violent local disturbances, and generating those sudden gusts and
squalls known as "Sumatras," which are so dangerous to navigation
in the Strait of Malacca.

Rivers.—Owing to the westerly position **of the** uplands, and a
general easterly tilt of the land, all the large watercourses necessarily
flow in the direction from west to east. On the west side the only
important river is the Singkel, which develops a winding course
through the Achin and Singkel districts to the north-west coast over
against the Banyak Islands. But on the opposite side a large number
of considerable streams make their way from the western ranges
across the lowlands eastwards to the China Sea. As the island
broadens perceptibly southwards between 2° N.—4° S. lat., so the
rivers increase in amplitude in the same direction, the largest being
the Siak and Indragiri respectively north and south of the equator,
the Jambi, about 2° S., and the Palembang (Musi), most voluminous
of all, between 3°—5° S. All are navigable for vessels of considerable
size, although somewhat obstructed by shoals, bars, and intricate
deltas about their lower course. Their numerous headwaters, converging from various points on the main stream, give them a fan-
shaped appearance, and cause extensive floodings of their low-lying
banks during the rainy season. The Siak, one of the most useful for
navigable purposes, is accessible to large vessels for 80, to ships of
200 tons for 100, and to boats for 150 miles from its mouth. But
the Palembang with its great tributaries, the Rawas on its left, the
Lamatang and Ogan on its right bank, presents a far more extensive
water system, navigated for some hundred miles, especially by large
bamboo "rakits," 40 feet and upwards in length. On one of these
rafts Mr. H. O. Forbes in 1881 floated down the Rawas from Pulu-

Kida, near its source, in 102° E. long., for over 200 miles all the way to the city of Palembang, still 50 miles from its mouth in Banca Strait. The trip occupied over three weeks in November and December during the wet season, when the banks of the main stream were flooded for some miles in many places to the great depth of 60 or 70 feet. A vast trade with the interior is carried on with these rafts, **which, like those on** the Tigris, are broken up and sold **for** their **valuable** material at their destination.

To its numerous eastern watercourses, combined perhaps with a slight upheaval of the land, Sumatra is indebted for its present ample dimensions. Originally probably not more than 100 miles broad, it expanded eastwards according as the mountain torrents encroached upon the China Sea, depositing the detritus of the Barisan highlands in its shallow waters, and thus gradually raising the marine bed above sea-level. In this way were created the great Sumatran alluvial plains, which for 200 miles inland are seldom over 400 feet high, and which are still constantly advancing seawards. The **time is** approaching when the 30 feet of water, now flowing between **the** east coast and the chain of eastern islands (Lingga, Sinkep, Banca), must be filled in, and then the east Sumatran seaboard will fall in a direct line with the southern extension of the Malay Peninsula. Thus the Asiatic mainland tends again to gather up its scattered insular fragments through the action of the Sumatran streams, which from short impetuous upland torrents have become great navigable arteries, winding sluggishly through the flat alluvial plains of their own creation.

Lakes.—Sumatra differs from most of the Malayan islands in **the** lacustrine character of its upland scenery. It possesses several romantic mountain lakes, the largest of which are *Tobah* in **the** Batta Country, about 3000 feet above sea-level, 20 miles long, source of the Singkel, and itself fed by numerous streams, chiefly from the north; *Singkara*, on the Padang plateau, 20 miles by 12 to 15, source of the Indragiri; *Danau Sapuloh Kotah*, or "Lake of the ten forests," at the foot of Mount Singalang in the north-west; *Korinchi*, in the Korinchi country, near the Indrapura Peak, unsurveyed; *Ranau* at the foot **of** the quiescent Sinmung volcano, in the Lampongs, 1700 feet above sea-level. This lake, visited in 1881 by H. O. Forbes, is of great depth, and teems with fish (one species, the semah or *Leobarbus*, as large **as** the largest salmon), which often perish in the hot water of the thermal springs of 127° F. bubbling up round the margin.

Climate.—The climate, especially **on** the uplands, is remarkably cool and salubrious. Frost, snow, and hail are unknown phenomena; but dense fogs and thunderstorms are of frequent occurrence. In the southern districts rainy days occur throughout the year, and in 1879 a rainfall of 282 inches was recorded at Padang.

Flora.—Sumatra being largely alluvial and volcanic, most of the soil is highly fertile, and suited for the growth of most tropical plants. Large tracts, however, are still held by rude tribes, who possess little knowledge of agriculture, depending for their existence nearly altogether on the spontaneous products of nature. The soil on the west side of the island is a stiff, reddish clay, while extensive districts, especially towards the south, are still under primeval forest.

Although still comparatively little known, the researches of Raffles, and more recently of von Rosenberg and H. O. Forbes, show that the Sumatran flora abounds in a great variety of tropical and sub-tropical species, on the whole more allied to those of Borneo than of Java or the mainland. Amongst the most useful are rice, sago, camphor, dammar, gutta, benzoin, dragon's blood, yielded by a species of rattan, bamboos, pepper, and tobacco. Peculiar to this region is the curious *Rafflesia Arnoldi*, discovered by Sir Stamford Raffles, a parasitic plant, with a flower over three feet in diameter, with very large brick-red petals, but possessing neither stem nor leaves, and simply adhering by minute fibres or roots to a species of vine. Other curious plants described by Forbes are a species of Sambucus, producing near its florets, little cups full of rich yellow honey, and the giant Arum (*Amorphophallus Titanum*), with enormous tubers over six feet round, and forming "a load for twelve men." But, notwithstanding the presence of *Melastomus* and some other beautiful flowering shrubs, the forest vegetation is remarkable rather for its bright-green, pink, or scarlet, and autumn-tinted foliage, than for its gay floral adornments.

Fauna.—The Sumatran fauna present far more numerous points of contact with those of the Malay Peninsula and Borneo than with the Javanese. Here are represented all the great mammalian forms of the mainland, such as the elephant, rhinoceros, and tiger, besides the tapir, the Siamang, a large ape, found elsewhere only in Malaya, and the Bornean orang-utan; this last confined to the wooded plains opposite Malacca. Of large domestic animals the most valuable is the buffalo, which as live stock takes the place of the European ox. There are several varieties of the monkey tribe, and Sumatra also possesses nearly all the beautiful and remarkable forms of birds common to Malacca and Borneo, besides a few species peculiar to itself. Very characteristic are the lovely Argus pheasant, which here takes the place of the peacock in Java; the *Rhotodytes diardi*, a species of cuckoo, with green bill and velvet scarlet eye-wattle; green and black barbets; scarlet Pieridæ, and the Buceros. Butterflies, beetles, and other insects are found in as great variety, perhaps, as in any other part of the Archipelago.

Minerals.—Sumatra is probably rich in minerals, as coal, naphtha, sulphur, iron, and gold have been found, as well as **indications** of copper. The ore is of fine quality, and the iron and **steel** produced in Menangkabau have been noted from time immemorial. Tin also exists, and is worked in Kampar, nearly opposite Malacca.

Trade.—The chief exports are black pepper, shipped in great quantities for Batavia and Singapore, maize, sago, cocoa-nuts, camphor, benzoin, dammar, cassia, cotton, gold-dust, and various tropical fruits. In exchange are taken rice, opium, salt, piece-goods, iron and copper ware, pottery, Chinese goods, dried fish, &c. The foreign trade of the country is carried on through the ports of Padang, Palembang, Bencûlen, Serdang, Deli, Muntok, Telok, Betong, Bengkalis, and Achin, recently taken by the Dutch. Steam communication is maintained between several of these ports and Singapore, Penang, Batavia, and other places in the Dutch colonies.

Inhabitants.—Apart from the still undetermined Indonesian element, both the cultured and uncivilised people belong to the Malay stock, which, under different names, and with varying languages, customs, and religions, is found diffused throughout the whole island. Hence many ethnologists have regarded Sumatra as the principal home of this race, and in any case it must be looked upon as the chief centre of dispersion for the civilised Malay people throughout the Archipelago during the last eight or ten centuries. No dark or woolly-haired race, analogous to the Negritos of Malaya and the Philippines, has ever been discovered in the island, which from the remotest times has been entirely occupied by Malay peoples, affected, especially in the north, by contact with immigrants from India, Arabia, and Indo-China. The result of these interminglings has been a considerable diversity of type and speech, as shown in the Achinese and Battas of the north, the Menangkâbaus of the west, and Jambis of the east central districts, the Sarawis, Palembangs, Rejongs, and Lampongs of the southern regions. Still more divergent types are those of the Kubus, Lubus, and other rude tribes scattered over the interior, many of **whom** should probably be grouped with the Indonesian family.

The Achinese of the extreme north, who have for many years maintained a war of independence against the Dutch, are regarded by Dr. Van Leent as true Malays with a mixture of Indian blood, presenting affinities with the natives of the Malabar and Coromandel coasts. Like so many other Sumatran peoples, they have developed a local culture, and have long been Mohammedans, writing their peculiar Malayan dialect with the Arabic characters. Their southern neighbours, the *Battas* (*Batta*, plural *Battak*), also continue to enjoy political autonomy, and are specially remarkable as the only known people who, although undoubted cannibals, possess a written language. They are a semi-civilised pagan nation, whose territory lies between Achin on the north, and the true Malay lands of Siak and Menengkâbau on the south. Their very peculiar culture seems to have had its earliest seat on the table-land of Lake Tobah, and was evidently at one time affected by Hindu influences, as shown by some Sanskrit elements in their language, and by the written character, obviously of Indian origin. Their cannibalism, which is of a mild form, appears to be a survival from still more ancient times, connected probably with the primitive rites of their rude ancestry.

Excluding the *Kubus* and some other little known wild tribes of the interior, all the other natives may be regarded as of pure Malay stock, with an admixture of Javanese blood, especially in Palembang. Those of Menangkâbau on the west coast appear to be the earliest distinctly Malay people who developed a national culture, and to this district many of the inhabitants of the Peninsula, of Borneo, and other parts of the Archipelago, directly trace their descent. The people of *Jambi* (Indragiri and Jambi basins), those of Palembang (Musi basin), the Rejongs further south, and the Lampungs in the extreme south, are also more or less civilised communities, possessing a knowledge of letters with several peculiar writing systems, and professing the Mohammedan religion. But, like their brethren in the Peninsula, they seem to attach more importance to the *âdat*, or national customs having the force of law, than to the precepts of the Qorán. Beneath an outward acceptance of Islâm, shown especially in their reluctant Friday attendance at the mosques, they cherish numerous superstitions surviving from Hindu and still older Pagan times, and a belief in the occult powers of nature is still universal. The people of Kissam in Palembang are even said to be still pagans (Forbes), and the richly decorated "Balai," or public assembly-room, conspicuous in every Palembang and Lampung village, is more highly venerated than the mosque itself. In the Passumah lands further north are some curious relics of a former culture, huge monoliths carved in the likeness of human figures, with strange non-Malay and non-Hindu features, although doubtfully ascribed by some to an Indian source. The present inhabitants, who would be quite incapable of executing such works, know nothing of their true origin, attributing them to certain mythical beings, who wandered over the land, turning all their enemies into stone. They are probably the work of the same forgotten race, who executed somewhat similar monuments in Easter Island and other places in the Pacific Ocean.

Political Divisions.—With the exception of Achin in the north, and the Batta territory south of it, the whole of Sumatra is under the direct or indirect control of the Dutch. All the southern section, as far as the Jambi river, and a broad tract along the west coast as far as Sinkel, in about 3° N. lat., are comprised within the residencies, or organised provinces dependent on the central govern-

L

ment of the "Dutch East Indies." North of the Jambi river are the native States of Jambi, Indragiri, Kampur, Siak, Assakan, Batu-Bara, Serdang, Deli, Langkat, and Riah, which acknowledge the suzerainty of Holland, while the Batta and Achin lands are still unreduced.

The Dutch possessions are administered by a Lieutenant-Governor resident at Padang, and under his jurisdiction are six separate Residencies, as under :

1. PADANG, on the west coast.
2. The PADANGSE-BOVENLANDEN, or "Padang Plateau."
3. TAPANULI, including Singkel, north-west coast.
4. BENCÚLEN, south-west coast.
5. LAMPÚNGS, the southern extremity of the island.
6. PALEMBANG, with Lambi, east coast.

Chief Towns.—*Padang*, capital of the west coast government, on the Padang river, the most important town in Sumatra, was founded in 1660, when the Portuguese were driven from a neighbouring factory. It is defended by a fort standing one mile up the river, and does a brisk export trade in pepper, camphor, benzoin, and coffee. Padang was first visited by the English in 1649, seized by them in 1781, and restored to Holland at the general peace of 1814. *Palembang*, next in importance, and a larger place than Padang, occupies both banks of the Musi, about fifty miles from its mouth, and accessible to large vessels. It is the great emporium of the inland trade, and has a large mixed population of Malays, Javanese, and Chinese, with some Dutch officials and soldiers. Here are the palaces occupied by the native princes before the kingdom of Palembang was reduced by the Dutch. *Siak*, a busy trading place in the north-east. *Bencúlen*, capital of a Dutch Residency, on the south-west coast, in an unhealthy district at the mouth of the Bencúlen river ; occupied by the English from 1685, when they quitted Batavia, till 1825, when they ceded it to the Dutch in exchange for Malacca. Although now a small place with an exposed roadstead, it still exports some pepper and camphor. Other seaports are *Achin*, *Deli*, *Muntok*, *Bengkalis*, *Telok Betong*, and *Tapanuli*.

ACHIN (properly ACHÉ, from the Hindustani *Achchá*, "good," "fine "), the northernmost town in Sumatra, gives its name to the independent Malay State, occupying the northern extremity of the island, whose extent is variously estimated at from 900 (Veth) to 1200 (Anderson) square miles. This State rose to great power under Sultan Iskander Muda (1607—1636), whose sway extended for 1100 miles round the coast from Aru opposite Malacca to Padang, and whose supremacy was also acknowledged by the island of Nias, and by the continental Malay States of Johor, Pahang,

Kedah, and **Pérak**. At present **its** limits are, on the east coast, the Tamiang, 4° **25′ N.**, separating it from Siak territory ; and on the west, the frontier of the petty State of Trumon, in 2° 48′ N. But Achin proper is now understood to be limited by a line drawn from Pedir Point on the north-east to Kwala Lambesi on the west coast. The inhabitants, who are akin to the neighbouring Battas, are chiefly occupied with the cultivation of rice, pepper, and betel, which they export in exchange for opium, salt, iron, and copper-ware, piece-goods, pottery, Chinese goods, dried fish, fire-arms, and ammunition. The capital lies close to Achin Head at the northern extremity, and west of the "Golden Mountain," a volcanic peak 6000 feet high. Achin was occupied by the Dutch in 1874, two years after the commencement of the hostilities which still continue, and which have for their object the complete reduction of this warlike and independent people. Since 1204 A.D. they have **been** zealous Mohammedans.

Historical Notes.—Of the ancient history of Sumatra **little is** known beyond the fact that many of the natives adopted some **form** of Hinduism at an early date, as is sufficiently attested by the Sanskrit elements present in their languages, and by some of the local monuments and inscriptions. The Javanese also seem to have largely influenced the southern populations, as shown by inscriptions in the Javanese character **occurring** as far north as Menangkábau, and by many purely Javanese names of places **both here** and in Palembang.

The most important subsequent event is the comparatively recent spread of Islâm throughout nearly the whole island. In the north it appears to have **been** adopted early in the thirteenth century, and in Marco Polo's time (about 1360) the people of the eastern seaboard were already followers of the Prophet. Menangkábau was the first Mussulman State that acquired political supremacy, and this district soon became the chief centre for the diffusion of the civilised Malay race and language throughout the Peninsula and Archipelago. All earlier migrations must be referred to Hindu and prehistoric times, the former from Java chiefly, the latter most probably from the Asiatic mainland.

Sumatra was first visited by the Portuguese in 1509, by the Dutch in 1600, and two years later by the English, who appeared at Achin with five merchant ships under Sir James Lancaster, bearer of a letter from Queen Elizabeth to the Sultan. The English continued to establish factories and settlements in the island during the seventeenth century, but principally in 1685-6. These settlements were retained till 1825, when they were all ceded to the Dutch in exchange for Malacca. Since that time the British have ceased to maintain any diplomatic relations with Sumatra, over which Holland now claims complete supremacy.

JAVA AND MADURA.

Sixth in size, and by far the richest and most populous of all **the East** Indian Islands, JAVA rivals the most favoured regions of **the** world in its fertility, natural beauties, and exuberant vegetation. **It** lies between 105° 10'—114° 34' E. long., and between 5° 52'—8° 46' S. lat., stretching from the Sunda Strait for 622 miles eastwards to the Bali Strait, with **an** extreme breadth of 121 miles from Cape Bugel in Yapara to the south coast of Jokjokarta, and an area of about 52,000 square miles. Both physically and administratively MADURA forms a dependency of Java, from which it is separated at its western extremity by the navigable Surabaya Strait, less than two miles wide. It is 96 miles by 18, and consists mostly of chalk, the cretaceous hills on the north side forming a geological continuation of those of Rembang and Surabaya on the opposite side of the strait. Hence **it** seems probable that it formed an integral part of the main**land** before the epoch of the great upheavals, of which Java was the **chief centre.** It has some extensive forests, but the soil is generally poor, yielding insufficient rice for the local consumption.

Other natural dependencies are *Pulau Panitan* (Prince's Island), lying off the westernmost promontory ; *Krakatau*, in the Sunda Strait, scene of the memorable eruption of August 26th, 1883 ; the small *Carimon Java* group, about 50 miles north of Yapara ; *Bawean*, some 60 miles due north of Surabaya Strait ; and *Deli, Tinjil, Nusa Kambangan, Sempu*, and *Nusa Barung*, off the south coast, making altogether a total area of nearly 52,000 square miles.

Coast-line.—Java, which is washed north and south by the Java Sea and Indian Ocean, has a coast-line of 1400 miles, diversified by several open bays on the north side, but with no deep inlets or natural havens, so that the only commodious harbour is that of Surabaya, formed by the adjacent island of Madura. The still less indented south coast has no safe anchorage except under the shelter of Kambangan, and is moreover exposed to the heavy and dangerous surge rolling in from the Indian Ocean throughout the year. The coast-line is otherwise broken by a number of bold headlands with intervening bays, such as Java Head and St. Nicholas Point at the southern and northern entrance of Sunda Strait: Wyncoop's, Welcome, and Pepper Bays at the west end of the island ; Cape Bugel over against Carimon Java, and Cape Sedano commanding the northern approach to Bali Strait.

Physical Features.—Java is traversed throughout its whole length by two mountain ranges, which in some places converge and again separate, throwing off numerous spurs, which fall gently down to the coast. Both are thickly set with about forty-six volcanoes, from 6000 to over 12,000 feet high, twelve of which are still active. The loftiest is Semeru (12,238 feet), but the largest crater is that of Tenger (8000 feet), which rivals in size those of Japan and Hawaii. The highlands are almost everywhere intersected by lovely valleys watered by torrents and perennial streams, and mostly overgrown with a magnificent tropical vegetation. There is, however, a singular absence of lacustrine basins, the only formations of this sort being a few small but romantic upland lakes in the provinces of Cheribon, Pasaruan, and one or two other places. The *Danau*, or "lake," in a pre-eminent sense, is now dry, like several others in different parts of the island.

Volcanoes.—Apart from the cretaceous and more recent alluvial deposits, the formation is essentially volcanic, Java forming, perhaps, the most important section of the great igneous zone, which traverses the whole Archipelago from Sumatra to the Philippines. But of all the larger islands it appears to be the poorest in useful minerals. Coal or lignite occurs in small "pockets" in many parts both of Java and Madura, as well as in the neighbouring islets; but all attempts have hitherto failed to utilise it to any great extent. A variety of clays suitable for bricks, earthenware, and porcelain; *ampo*, an edible earth, regarded as a delicacy by the natives; good limestone and marble, petroleum and sulphur, abound in many places, while salt is obtained from the mud wells of Kudwu and Selo (Samarang), and saltpetre from Sutyi in the Gresik district.

Thermal wells also abound, and the volcanoes yield at every eruption large quantities of sulphur. The crater of Tashem, at the east end, contains a lakelet one-fourth of a mile long strongly impregnated with sulphuric acid, whence flows a stream of acid water so destructive to life that no fish can live in the sea near its mouth. The widespread myth of the deadly "upas tree" originated probably with the extinct volcano of Guevo Upas ("Vale of Poison"), in the Batur district, whose crater, about half a mile round, is justly held in horror by the natives. All living things approaching it drop down dead, and the ground is strewn all round with the remains of deer, birds, and even men, killed by the carbonic acid gas rising from the bottom of the valley. The mud volcanoes of the low-lying Grobogan district in Yapara yield a considerable supply of the muriate of soda useful for culinary purposes. In the neighbourhood of the Bromo (Tengger) volcano the fire used for cooking is always taken from the incandescent matter ejected by that venerated mountain during eruptions. This fire is religiously preserved for years, and whenever it goes out is kindled anew from that of the nearest village obtained originally from the volcano. The fires in use up to the late outburst were all procured from the Bromo eruption of 1832.—*Straits Times*, 1886.

Earthquakes are frequent, but seldom violent, and mostly of a local character. Nor do they now appear to have any intimate connection with the eruptions ; of 143 recorded by Junghuhn, not more than 24 have been in any way associated with these disturbances. But the memory survives of tremendous convulsions formerly accompanying the explosions, such as that of Ringghit in 1586, when that giant was rent asunder, involving in the ruins 10,000 persons ; and that of Gahing-gung in 1822, which swept away 115 villages with their 4000 inhabitants.

Rivers.—The northern versant of Java differs from the southern in the great development of its alluvial formation, which in some places forms plains of considerable extent. Hence the streams flowing to the Java Sea greatly exceed in length and volume those falling into the Indian Ocean, none of which are navigable for large vessels. Most of the tyi, or "rivers," as they are called in the western districts, are, moreover, obstructed by shoals or sandbanks at their mouths. The largest, and in some respects the most useful, is the Bengawan, or Solo, which flows by the city of Surakarta for 350 miles eastwards to Surabaya Strait, and is navigable for large boats. Next in magnitude is the Brantas, called in its lower course the Kalimas, and by Europeans the river of Surabaya, which after a winding course round Mount Arjuna, falls through two mouths into the same strait. The rapid formation of alluvial deposits at their mouths gives abundant proof of their disintegrating agency. Similar accretions of land are taking place all along the north coast, where steam-dredges have to be kept at work in all the large harbours.

Owing to their generally rapid course and perennial character, the innumerable streams and torrents on both slopes of the island are, on the whole, far more useful for irrigation than for navigation. To the extensive practice of this art, combined with the rich character of the volcanic and alluvial soil, are mainly due the magnificent crops which enable Java to support considerably more than half the population of the whole Archipelago.

Climate.—The Javanese year is divided into a wet and a dry season, the former lasting from October till March, when the moist westerly winds prevail, the latter for the rest of the year, when the cloudless east monsoon predominates. The driest months are July and August, when the days are hottest and the nights coolest. At Batavia the glass ranges from 70°—74° F. in the morning, to about **83°**, and occasionally even 90° at noon. But on the uplands of the interior the extremes lie between 60° and 70°, falling to 50° and 62° on the hills behind Samarang, and to 27° or 5° below freezing point on the summit of Mount Sindoro. On all the highlands the climate may be regarded as healthy for Europeans, who become enervated in the hot and often insalubrious low-lying districts along the northern seaboard.

Flora.—Rice, the staple of food for all classes, is grown not only along all the coast-lands, but on all the lowlands and valleys where water is available. It is replaced on the uplands by coffee, which has become the chief article of export. During the ten years ending 1878 the average annual produce of the Government plantations was 878,000, that of the private planters, 156,000 piculs. The export of coffee is entirely in the hands of the "Netherlands Trading Society." Other vegetable products are sugar, raised chiefly in the Batavia district where numerous sugar-mills are now at work; tobacco, maize, pepper, cinnamon, cloves, nutmeg, sago, indigo, tea, and pimento. Palms and cocoa-nut trees abound in great variety, and are distinguished by their luxuriant growth, sometimes reaching the height of 150 feet. Fruits of exquisite flavour, such as the mangosteen, durian, rambutan, mango, plantain, guava, pine-apple, are largely grown, and of late years the cinchona has been successfully cultivated by the Government, which now possesses 1,680,000 trees. The cultivation of tea, begun by Du Bus, has also attained a considerable development, the production amounting in 1879 to over 5,700,000 lbs.

A characteristic forest plant is the far-famed *Upas*, that is, "Poison," whose sap is fatal to all animal life. Extensive forests of *jati* (teak) occur, especially between Samarang and Sidayu, and yield a timber of finer quality than that of Burmah. The spices thrive well, but are not much cultivated, and the vine, formerly extensively grown, was stopped by the old Dutch East India Company for fear of prejudicing the South African vineyards. In the central and western forests are found many valuable trees, including as many as sixteen varieties of the oak. But the woodlands are everywhere exposed to two destructive agencies—the *alang alang* cane, an ineradicable exhauster of the soil, highly injurious to all other vegetation; and the upland peasantry, who clear the land for tillage in the most reckless manner. On the lowlands a better method of cultivation prevails, known as the "culture system," introduced by Governor-General Van den Bosch over 50 years ago. Under this system the great staples of agriculture have increased wonderfully, and although scarcely more than one-third of the land is under cultivation, Java now produces not only enough grain for its own teeming populations, but has also become a chief source of supply for the whole Archipelago.

Fauna.—The domestic animals are the horse, cattle, sheep, goats, swine, and buffaloes, the last-mentioned being almost exclusively employed in field operations. The Javanese fauna is in other respects mainly Asiatic, including the tiger, panther, leopard, jackal, polecat (*mydaus melicps*), rhinoceros, wild ox, deer, two species of wild boar, the wau-wau gibbon, and other members of the ape family. But the elephant and tapir are absent, or have long been extinct, if their range ever extended so far south. Of reptiles the most formidable are the crocodile and python, both numerous and of large size, besides upwards of twenty venomous snakes. The ornithology is very rich, including the cassowary, peacock, weever, two species of parrot, the minute butterfly hawk, falcon, golden oriole, yellow crowned bulbul, fairy blue-bird, jungle-fowl, and many other rare and beautiful species. The rivers also and neighbouring seas are well stocked, and the fisheries along the coasts highly productive. But the rivers' mouths are infested by alligators, and the surrounding waters by still more voracious sharks.

Inhabitants.—All the natives belong to the Malay stock, speaking three distinct but allied languages: *Sundanese* in the west; *Javanese* proper in the central and eastern provinces; *Madurese* in Madura and adjacent parts of the larger island. . In physical appearance they present little differences, except that the Javanese are somewhat taller and perhaps more refined than most other branches of the Malay race. All are naturally inoffensive, peaceful, docile, of frugal habits, truthful and straightforward. They bear the impress of a people that has long enjoyed the benefits of a stable government, of social order, and a considerable degree of general culture. Their husbandry is careful and orderly, and they betray much skill and taste as workers in wood, iron, and other metals. Their boats and canoes are unsurpassed for speed and elegance, their krisses of excellent temper and graceful design, their woven fabrics of fine quality, with tasteful patterns and harmoniously blended colours, derived from a few simple vegetable and other dyes. As musicians they have always excelled amongst Malay peoples, with whom, however, they share the love of gambling, of cock-fighting, and some other characteristic vices.

The Javanese language, current in the greater part of the island, is derived directly from the *Kavi*, a highly developed form of Malay speech, of which there are inscriptions and records dating from the twelfth century. It is written in a peculiarly elegant syllabic character, which was introduced in an older form from India, and which has held its ground even after the Hindu religions were supplanted by Islám in the fifteenth century. Since that time the bulk of the people are reputed Mohammedans, although really believers in the primitive animism of their forefathers.

Many Brahmanical and Buddhist ideas also survive, and the number of *hyangs*, or spirits, still worshipped is limitless. Every village has its patron deity, who presided at its foundation, and to whose beneficent or malignant influence are ascribed all its fortunes. Under a broad-branching tree stands the altar, on which the worshipper lays his offering of flowers and incense, uttering meanwhile in broken Arabic the Moslem formula—"There is no god but God, and Mohammed is his prophet." The national pantheon has also been increased by such names as Moses, Jesus, St. Joseph, and others, introduced through European influences, although their Protestant rulers have hitherto done little to evangelise these "Mohammedan and Hindu Nature worshippers." More zeal has been shown by the Roman Catholic Church, which has a Vicar-Apostolic resident in Batavia, and subordinate to him many missionaries scattered over Java and other parts of Netherlands India. At Batavia and Samarang there are religious establishments for the education of the young on Christian principles. A spark of the old Hindu religions is still kept alive by the Jeins (Badui) hill-men in the Lebah district, Bantam, and by the more numerous Tengger people, who occupy the slopes of the Tengger volcano.

Government.—The only native princes still retaining a semblance of regal state are the rajas of Surakarta and Jokjokarta, who are mere pensioners of the Dutch, with no power to levy taxes, but with absolute jurisdiction in purely religious matters. For all practical purposes Java and its dependencies are now directly administered by the Dutch, who have parcelled out this region into twenty-four Residencies, enumerated in the statistical tables, p. 187. These Residencies, each of which is governed by a European Resident, assisted by a secretary and a number of sub-residents, are subdivided into arrondissements, or "regencies," so called because entrusted, especially in police matters, to native chiefs named "regents." Over all stands the Governor-General, who resides at Batavia, and exercises almost absolute authority over all the Dutch possessions in the Archipelago. He is Commander-in-Chief of the land and sea forces, and is assisted by a Secretary-General and a Colonial Council of four members named by the King of Holland.

Trade.—Java is the centre of a large and increasing local and foreign trade, which has been greatly facilitated by an excellent system of roads, and a network of railways connecting all the chief towns along the north coast with each other, and with several points of the interior. Regular lines of steamers also ply between Batavia and Europe, Singapore, Padang, and all the chief ports of the Archipelago. The exports from Java include rice, sugar, coffee, indigo, tobacco, cotton, pepper, spices, camphor, teak, sago, and edible birds'-nests. Through Batavia are also forwarded to Europe many other products of the Archipelago, such as gambier, tin, gold-dust, diamonds, rattans, beeswax, tortoise-shell, nutmegs, cloves, mace, kajuputi, and other oils from the Moluccas. The total yearly exports now exceed sixteen, and the imports thirteen, millions sterling. The latter comprise linen, woollen, and cotton goods, provisions, wine, spirits, hardware, glass from Europe and America; opium from India; tea, porcelain, and silks from China. All the Government exports to the Netherlands are forwarded by the "Dutch Trading Company" established in 1824 at Amsterdam.

Topography.—*Batavia*, capital of Java, and of all the Dutch East Indies, occupies a marshy site on the Yakatra, near the head of the spacious Bay of Batavia. It comprises a native and a European quarter, the latter rivalling Calcutta and Bombay in splendour, and containing the residences of all the Government officials, all the chief hotels, clubs, museums, and theatres. But business is centred chiefly in the old town, which is intersected by canals, and rendered as salubrious as most tropical cities by the recent drainage works. Nearly all the import and export trade of Java passes through Batavia, which has a mixed population of over half a million natives, Chinese, "Moors" claiming Arab descent, Dutch, English, Portuguese, and other Europeans. It was founded in 1619, and occupied by the British in 1811, but restored to Holland at the general peace.

About 40 miles south of Batavia, in a healthy district nearly 1000 feet above the sea, lies the village of *Buitenzorg*, where the Governor-General has a fine palace, and many Europeans reside a part of the year. Here is a famous botanical garden, in which are cultivated all the finest vegetable products of the Archipelago. *Surabaya*, next in importance to Batavia, and the chief port for the export of sugar, stands at the mouth of the Brantas river, over against the western extremity of Madura. Its harbour is the finest in Java, and here are situated the Government dockyards and arsenals. The fertile province of Surabaya sends down a vast quantity of rice, sugar, and other produce by the river, which is navigable for large boats far into the interior. *Samarang*, at the mouth of the river of like name, some 220 miles east of Batavia, enjoys the advantage of railway communication with the native capitals of Surakarta and Jogjokarta, thus drawing large supplies of cotton, sugar, coffee, and indigo from one of the richest districts of the interior. *Anjer*, a fortified town at the narrowest part of the Sunda Strait, and an important port of call for ships proceeding to Batavia, Singapore, or Manilla, was totally destroyed by the terrible Krakatao eruption of August 26-7, 1883. *Surakarta* (Solo), capital of the kingdom of Susuman, the so-called "Emperor of Java," is the largest city still nominally governed by a native prince. He keeps a ceremonial state, and is surrounded by a degree of magnificence scarcely surpassed by that of any Indian raja. Another important native city is *Jokjokarta*, capital of a province of like name, and also governed by a Javanese sultan.

Historical Notes.—Like that of India, the early history of Java lacks all satisfactory chronological data. The first known records, as well as the oldest monuments, are associated with the ascendancy of the Hindus through three successive periods of pure Buddhism, an aggressive Sivaism, and an apparent compromise. Of the various Hindu States the most powerful was that of Majapahit, with many tributaries in Java and other parts of the Archipelago. Under Buddhist and Brahmanical influences the peaceful arts, and especially architecture and sculpture, attained a degree of almost unparalleled splendour, as still attested by the sumptuous monuments of Boro-budor and other places. But Hinduism was almost entirely displaced in the fifteenth century by Islâm, which, as a political power, had in its turn soon to give way to Christian influences.

Java was first visited in 1511 by the Portuguese, who were followed in 1595 by the Dutch. For over a hundred years the Dutch East India Company owned only a few forts and factories at Yakatra (Batavia) and other places; but in 1705 they obtained possession of Preanger by treaty with Mataram, and in 1745 extended their authority over the whole north coast from Cheribon to Banyuwangi. In 1755 Mataram was divided into the two States of Surakarta and Jokjokarta, which still retain a semblance of independence, and in 1808 the kingdom of Bantam was finally reduced. By the British occupation (1811-18) European ascendancy was strengthened, and the great Java war (1825-30), in which a last struggle was made by a native dynasty, resulted in the complete triumph of the Dutch. Since then the whole island has fallen under their sway, and under their able administration has rapidly increased in population and general prosperity. A great military road 600 miles long and other highways have been opened in every direction, the railway system is gradually extending to all the great agricultural centres, life and property are as safe as in any part of Europe, and the natives are as contented as any people are likely to be under the rule of an alien race.

Of the numerous monuments left by the early Hindu conquerors, the most remarkable is the great temple of Borobudo (Boro-Budur), about 80 miles west of Brambanam, in the province of Kedu. It crowns a small conical hill, and consists of a lofty central dome, and seven ranges of terraced open galleries, regularly built round the hill, with numerous communications by steps and stairs. The dome is 50 feet in diameter, and the whole building 620 feet square and about 100 feet high, representing an amount of labour as great as that expended on the Great Pyramid. Of minor antiquities the most valuable are the inscriptions on stone and

copper in a variety of characters, rendering their decipherment a work of
great difficulty.   Some of these relics appear to have been removed to the
Raffles Museum, Singapore.

BALI.

Last of the lands belonging physically to the Asiatic division of
the Archipelago, this westernmost of the Lesser Sunda group lies
between the shallow Bali Strait now separating it from Java and the
deep Lombok passage, by which it has always been severed from
Lombok and the whole Australasian world.   It has a circumference
of about 200 miles, and is mostly hilly and even mountainous, culminating in the north-east with the volcanic Gunong Agung, or Peak
of Bali (11,400 feet).   From this and other hills flow numerous
streams in all directions, supplying abundant water to its fertile soil,
which yields rich crops of rice, cotton, and tobacco.   The natives,
akin to the Javanese in type and speech, are a finer and a more independent race than their neighbours, as shown by their stout resistance
to the Mohammedan invasion.   Here the Hindu forms of religion
have found a last refuge in the Archipelago, most of the people being
sectaries either of Brahmanism or Sivaism, as in India itself.   There
are also a few Buddhist communities, but scarcely any followers of
the Prophet except amongst the Malays of the trading places.   The
institution of castes even still prevails, and *satti*, or the immola-
tion of widows on their husband's funeral pyre, has not yet been
suppressed.

The island is divided into eight principalities (Beleling, Karang-asam,
Klong-kong, Tabanan, Bangli, Maugiri, Gyangar, and Badong), whose
hereditary rulers retain the title of raja.   But since 1849, when Bali was
completely reduced by the Dutch, these potentates enjoy the mere semblance of authority, and the island now forms with Lombok a Residency,
administered by an official stationed at *Beleling* on the north coast, the
chief seaport of Bali.

THE PHILIPPINE AND SULU ARCHIPELAGOES.

General Survey.—The Philippines occupy the whole of the
north-eastern section of the Archipelago, with the central parts of
which they are physically connected by three chains of islands—
Palawan with Balabac, running from Mindoro to north-western
Borneo, the Sulu group from Mindanao to north-eastern Borneo, and
Carcarolong with the Talautse (Sanguir) and Siao groups, also from
Mindanao to Celebes.   These chains all lie on more or less elevated

marine beds, enclosing the two deep basins of the Sulu and Celebes
Seas, while further north flow the deep waters of the China Sea, now
completely severing the Philippines from the Asiatic mainland.
Again, Palawan and Sulu appear to consist mainly of very old
sedimentary rocks, while Talautse and Siao are exclusively volcanic,
the Philippines themselves partaking of both formations in more
equal proportions than any other section of the Eastern Archipelago.
This twofold aspect, partly oceanic, partly Asiatic, is also presented
by their fauna, flora, and inhabitants, which, moreover, offer many
peculiarities, distinguishing this region from all others in the eastern
seas. Stretching north and south across 15 degrees of latitude
(5°—20° N.), with a total area of 115,000 square miles, it forms, next
to Great Britain and Japan, the largest compact insular group in the
world; and so closely are its various members connected, that they
produce the impression of a continuous mass of land broken into
fragments by the convulsions and subsidence so often associated with
igneous disturbances. Thus all the broad features here indicated
seem to point at one conclusion—that the Philippines represent a
vast area at one time contiguous with the continent and with Borneo,
then at a very remote period severed from both, and again partly
united with the Oceanic world through the more recent volcanic
agencies, of which Sanguir and Siao have long been an active scene.

The group comprises, 1. the two great islands of *Luzon* and *Mindanao*
in the north and south, the former somewhat larger than Java, the latter
one-fifth smaller; 2. the intervening islands of *Mindoro, Panay, Negros,
Zebu, Bohol, Leyte, Masbate,* and *Samar,* ranging from 1200 to 6000 square
miles in extent; 3. the outlying Palawan with the Calamians and Balabac
groups between the Mindoro and Balabac Straits; 4. the Babuyan, Catan-
duanes, Surigao, and other smaller groups, making altogether over 400 inhabited isles and islets, besides some 600 uninhabited islets and reefs.

Physical Features.—Throughout its whole extent the Archipelago **seems** to be traversed by two **somewhat parallel volcanic**
zones, but gradually converging southwards, so that the space of over
100 miles separating them in Luzon is reduced in Mindanao to 55
miles, while both merge in one system towards Sanguir and Siao. **Of**
the western and less elevated zone the chief cone is that of Taal in
the province of Batangas, rising 530 feet above lake Bombon and
1600 above sea-level. In the eastern zone the most important range
is that of Mayon, terminating at Point Engaño in the extreme north-
east of Luzon, and reappearing in the Babuyan isles. Southwards
this range culminates in Mount Albay, at the south end of Luzon,

passing thence through Leyte, where there are large deposits of sulphur, to the volcanic islet of Camiguin off the north coast of Mindanao, and so on to Apo and the Sarangani mountain and islets at the southern extremity of Mindanao.

Albay **is one** of the most remarkable volcanoes in the whole world, forming a regular cone 9100 feet high, with a circular base 12 miles in diameter, constantly emitting from its flanks thousands of jets of heated sulphurous vapour, but without a trace of any true crater. But during the eruptions of 1767 and 1814 it ejected torrents of lava which swept away many villages with all their inhabitants. Further north the volcanic region is sharply limited by the course of the river Bicol, south of which nothing occurs except calcareous marls and rich fossiliferous deposits. Here the Mayon system is continued north-westwards through Mounts Iriga (4000 feet) and Isarog (6500 feet), whose eruptions appear to have filled in the channel between the former island of Caramuan and the province of South Camarines. In North Luzon the eastern and western volcanic belts, which here enclose the Tajo river basin, take respectively the general names of the Sierra Madre and Northern Cordilleras. In Mindanao the still active Mount Apo, near Davao, was ascended in 1882 by Koch and Schadenberg, who found the highest of its three peaks to be 11,000 **feet**, consequently the culminating point of the whole Archipelago. The more southern Sarangani has been quiescent since 1645.

The presence of very old crystalline rocks in both of the large islands is attested by the occurrence of gold in Mindanao, and of auriferous quartz, lead and copper ores in the southern districts of Luzon. Similar formations occur in Masbate, whose streams are washed for gold, in Zebu, Leyte, and other members of the Archipelago, where igneous and sedimentary rocks are found almost everywhere intermingled.

Rivers and Lakes.—Few tropical lands are better watered than the Philippines, which, besides innumerable perennial streams, also differ from Java and Borneo in the possession of several large and romantic upland and lowland lakes. Of the Luzon rivers, which flow mostly to the north and west coast, the largest is the Tajo (Aparri), which flows from Mount Lagsig for about 200 miles through the great Cagayan plain to Aparri on the north coast. Further south the still larger Pampanga plain is traversed by a large number of streams, flowing some to the Gulf of Lingayan, some to Manila Bay on the west coast. Here is presented the somewhat rare phenomenon of a lake, the Laguna de Canaren, draining in two opposite directions to both of these inlets. Numerous streams also converge from the surrounding hills in the beautiful Lago de Bay, largest of the Luzon lakes, which sends its overflow through the Pasig river to Manila Bay near the capital.

Other large sheets of water in Luzon are Lake Cagayan in the extreme north, the temporary Pinag de Caudava, formed during the rainy season by the overflow of the Pampanga, and Bombon, a large lake 15 miles by 10, out of which rises the Taal volcano, whose deep crater is itself flooded by a lakelet three miles in circumference. The greater part of Mindanao is drained by two large rivers—the Butuan, flowing from near Mount Calalan northwards to Butuan Gulf, and the Rio Grande, which traverses a series of lakes on its westerly course to Illana Bay. Between the two, and in the very heart of the island, lies the extensive but little known lake Maguindanao, which, like the Laguna de Canaren, is also said to discharge its surplus waters in opposite directions to both of these river basins. Several other lakes are dotted over the interior of Mindanao, the largest of which appears to be Malanao, draining to Iligan Bay, on the north coast.

Climate.—Three seasons are distinguished, at least in the northern section of the Archipelago, which alone is exposed to the terrific typhoons that sweep with such destructive force over the China Sea. The cold and dry season, ushered in by the north-east monsoon in November, is followed by the secar, or period of heat and drought prevailing from March till June, when the heat on the lowlands is sometimes almost unbearable. The third, or rainy period, accompanying the south-west monsoon, prevails generally from June till October, when the typhoons are most frequent and violent. But they extend no further south than about 10° N. latitude, so that Mindanao and the Sulu group lie beyond the range of their devastations. The rainfall exceeds 100 inches in many places, and as most of the moisture is precipitated during the wet period, the lowlands are periodically flooded by the tremendous downpours of the summer and autumn months. In the north the greatest heats appear to prevail from April to July, or August, when the glass rises to 96° or even 100° F., while at other times falling to 75° and 72°. Owing to the absence of storms, the climate is more equable in the south, so that the distinctions between the seasons are much less perceptible in Mindanao than in Luzon.

Flora.—With the progress of exploration the number of indigenous forms is constantly increased. Thus the 2729 species and 910 genera already recorded in 1879 had risen in 1883 to 4583 species and 1163 genera. Most of the latter are common to Malacca, Borneo, and other parts of the Indo-Malayan region, while some belong to the Australasian world, and a few are indigenous. But owing to their long isolation, the Philippines have developed a greater number of species and varieties than any other insular group in the Archipelago.

The splendour of the highland scenery, which all travellers describe in enthusiastic language, is largely due to the magnificent forests of ebony, ironwood, cedar, sapan, teak, and many other valuable species, clothing all the slopes to a height of some thousand feet. In general the vegetation from 5000 feet upwards is identical or closely analogous to that of Borneo at the same altitude. Conspicuous amongst the cultivated plants are the sugar-cane, of which over twenty varieties are enumerated, tobacco, rice, hemp, coffee, all of excellent quality and great economic value. The bamboo, especially the so-called *Cauayang-tatoo* variety, is also of great importance for the endless social and industrial uses to which this indispensable graminaceous plant is put.

Fauna.—The Philippine **fauna is** remarkable especially for the total absence of rhinoceros, elephant, tiger, tapir, and all the larger **animals** common to other parts of the Indo-Malayan world. On the other hand, amongst the smaller forms special types are met in constantly increasing numbers. These indigenous varieties are in fact numerous enough to impart a peculiar stamp to the local fauna (*Jordana y Morera*). Thus the presence of many mammals akin to those of the adjacent lands shows that the isolation of the Archipelago cannot date from extremely remote geological **times,** while the absence of others of the same group may be due to the devastations caused by the tremendous volcanic and seismatic convulsions, as well as by the subsidence and upheavals, of which these islands have always been a chief centre.

Characteristic animals are the *macacus cynomolgus*, a species of ape spread over the whole group, a small panther, confined to Palawan, a wild cat, a mouse-deer, and flying mammals, which are exceptionally numerous, including a squirrel, a lemur, and over twenty species of bats. Many kinds of birds common to other parts of Malaya are also wanting, and partly replaced by a large variety of parrots and pigeons, besides cockatoos and mound-builders. The reptile class includes crocodiles, lizards, an enormous python over 40 feet in length, and many other snakes. "Some of the butterflies are remarkable for their intense and variable metallic gloss, and the Philippines are celebrated above all other eastern countries for the variety and beauty of their land-shells, of which there are about 400 distinct species, of varied form, and often of exquisitely delicate colouration" (*Wallace*).

Inhabitants.—Excluding the already described few surviving *Aetas*, or Negrito aborigines (see p. 120), the whole of the native population belongs fundamentally to the Malay stock, which, however, here presents a far greater variety of type and speech than in any other Malay region proper. Besides the larger nations, such as the *Tagalas* of Luzon and Mindoro, the *Bisayans* widely diffused throughout the central islands, the *Bicols* of South Luzon **and** Masbate, the *Mandayas* and *Manobos* of Mindanao, there **are many**

other tribes, especially in Luzon, who differ greatly in physical
appearance, usages, language, and general culture. Thus while the
Tagalas and Bisayans possessed a knowledge of letters, and rose to
a certain degree of civilisation in comparatively remote times, the
cruel and ferocious *Ilongotes* of the Caraballo highlands, Luzon,
are described as the most degraded of beings, destitute of all the
finer sentiments of humanity, and incapable of any generous act.

In Luzon there are no less than fifteen, and in the whole Archipelago some
forty languages, often presenting profound differences, although ultimately
reducible to the common Malayo-Polynesian stock. The cultivated Tagala
and Bisayan are far more highly developed than either the Malay or
Polynesian proper, and have evolved many curious and intricate forms of
speech, which seem to place them in an intermediate state between the
agglutinating and inflecting linguistic families. This surprising diversity
of types and languages must be attributed partly to the long isolation of
the Philippines, attested also by their animal and vegetable forms, partly
to the frequent contact with Asiatics and other peoples to which this
group has been exposed from prehistoric times, and partly to intermixture
with the Negritoes already in possession of the whole Archipelago before
the arrival of the first Malay intruders. The endless variety of races
resulting from all these causes is well illustrated by the 'Album of
Philippine Types' issued in 1885 by Dr. A. B. Meyer of Dresden, and
containing about 250 figures, showing every shade of transition from the
pure and half-caste Negrito and Malay to the Hispano-Malayan Mestizo.

A peculiarity of these populations is the resistance they have offered to
the spread of Islâm, contrasting in this respect with their susceptibility to
Christian influences. Of the total population about five and a half millions
are classed as "reduced," the majority of whom are members of the Roman
Catholic Church subject to the Spanish Government, leaving scarcely more
than 600,000 Negritoes, Chinese, and *Igorrotes*, a term commonly applied
collectively to the pagan and uncivilised Malay tribes, in contradistinction
to the *Ilotes*, or native Christians. The Christianity, however, of these
Ilotes is often purely formal, a mere outward cloak, beneath which heathen
rites and the lower phases of Romanism meet as on common ground.

These remarks do not apply to the Sûlûs of the Sûlû Archipelago, who
are of purer Malay descent, although also to some extent affected by
Chinese and perhaps Arab elements. All are zealous Mohammedans, and
were till recently notorious corsairs, a terror to the more peaceful seafaring
populations of the China Sea. But since the reduction of this group by
the Spaniards in 1876, their piratical expeditions have almost entirely
ceased, while the Sûlûs long settled in North Borneo have mostly become
orderly British subjects, somewhat indolent and restless, but on the whole
" well behaved, courteous, and intelligent" (W. B. Pryer). Palawan
(Paragua) also is partly inhabited by heathen Malays, the Tagbanuas, and
Tinitianos, who have many features in common with the Mohammedan
Malays of Mindanao. Farther north and in the neighbouring Calamianes
group live the Bulalacaunos, a ruder people, whose aquiline nose, crisp or
wavy hair, and somewhat full beard, separate them from the Malay stock,
and affiliate them probably to the Galelas and other Indonesians of the

M

Eastern Archipelago. Palawan is only nominally under the authority of the Spaniards, who maintain a solitary military establishment at Porto Princesa on the east coast.

Government.—For administrative purposes the Archipelago is divided into forty-three departments or provinces, governed by Alcaldes or Commandants, under the general control of a Governor-General and Captain-General. These officials are practically absolute in their respective jurisdictions, the great bulk of the population being still unripe for civil or communal rights. A large degree of authority is also beneficially exercised by the ministers of the Roman Catholic Church, which here boasts of more numerous congregations than in any other part of the Asiatic world. The hierarchy comprises an archbishop (Manila), three bishops, and nearly 500 parish priests, supported by a small poll-tax levied on all Christians, and by the revenue of large landed estates. The public revenue is also derived to a great extent from a capitation tax, supplemented with custom-dues, a tobacco monopoly, an excise on palm-wine, and a few other sources.

Trade.—In the absence of railways, or even good roads and bridges, the natural resources of the Archipelago still remain to a large extent undeveloped. Nevertheless, a considerable export trade is supported by the produce of the sugar, tobacco, hemp, and coffee plantations. Cotton and rice are mostly required for the local consumption, but other articles of export are timber, especially sapan, indigo, gums, hides, and mother-of-pearl. The chief imports are cottons, hard-ware, crockery, China goods, and provisions of all sorts.

Topography.—Nearly all the foreign trade of the Archipelago passes through the capital, *Manila*, founded in 1571 near the mouth of the river Pasig, on the east side of Manila Bay, west coast of Luzon. Manila has extensive commercial relations with China, Europe, and America, and enjoys telegraphic communication with the rest of the world through the cable to Hong-kong, laid down in 1880. As the centre of the Spanish possessions, and next to Goa, the oldest European town in the East, Manila possesses numerous public buildings, scientific, charitable, and religious institutions. The climate, although hot (mean temperature 82° F.), is not unhealthy ; but the place is exposed to terrific hurricanes (typhoons), thunderstorms, and earthquakes.

Next in importance to Manila is the seaport of *Iloilo*, on the south coast of Panay, the outlet of the best hemp-growing district. Other small centres of trade and population are *Zebu*, on the east side of the island of

like name, where Magellan was killed in 1521; *Zamboanga*, at the western
extremity of Mindanao; *Suál*, on the west, and *Aparri* on the north
coast of Luzon.

SÚLÚ.

This Archipelago, stretching for 200 miles in a gentle curve from
Mindanao to the north-east extremity of Borneo, comprises the three
groups of *Basilan* in the east, *Súlú* proper in the centre, and *Tawi-
Tawi* in the west. Until 1876 the whole chain formed an independent State under a Mohammedan sultan; but in that year the
Spaniards occupied Basilan, incorporating it with the Philippines,
and at the same time compelled Sultan Mohammed-Yamalal-Alam
to accept their protectorate. He now resides at *Maibun*, his former
capital, *Tianggi*, having been destroyed by the Spaniards, and since
replaced by a new station in a more healthy site. The district of
Súlú claimed by him in North Borneo now forms part of the British
settlement in that region, his territory being thus reduced to the
insular groups of Súlú and Tawi-Tawi. The two largest islands,
both about 35 miles long, and from 3 to 10 broad, are fertile and
partly covered with teak, sapan, cocoa and areca palms. Tobacco,
recently introduced by the German Borneo Company, thriving well.
(Guillemard.)

BASHI.

To the Philippines also belongs politically the small *Bashi*, or
Batanes, group between Babuyan and Formosa. Discovered by
Dampier in 1687, it was occupied in 1783 by the Spaniards, who
maintain a small establishment on Grafton islet. There is a considerable Malay population, who cultivate yams, batatas, pineapples,
and other fruits of fine flavour. The Dominicans have a station on
Batan, a high pyramidal island, which gives an alternative name to
the whole group.

II. OCEANIC DIVISION.

CELEBES—THE MOLUCCA AND BANDA GROUPS.

The reasons for constituting this region, which fills up nearly
the whole space between Borneo and New Guinea, a separate division
of the Eastern Archipelago, distinct both from the Asiatic and

Australian sections, are **fully** set forth at pp. 109, 110, **and** need not
here be repeated. Whether the lands of which **it is** composed be
regarded as a remnant of the vanished Lemuria, formerly stretching
south-westwards to Madagascar, or of a submerged Pacific continent,
which, with the Philippines, extended eastwards and south-eastwards
to the Marshall group **and** New Zealand (for this theory has also
been advocated), it seems evident that Celébes and the Moluccas
can have formed no part of the Asiatic **or** Australian mainland, at
least since Miocene times.

CELÉBES.

General Survey.—Lying almost exactly in the centre **of the**
Archipelago, of which it is the fourth largest member, **exceeding in**
size both Luzon and Java, Celébes stretches two degrees **north and**
nearly six south of the equator, between 119° and 125° E. **longitude.**
This strangely-shaped island, roughly resembling a starfish that has
lost one of its rays, consists of a central nucleus, whence radiate **north,**
east, and south four great limbs, traversed by four mountain-ranges,
and enclosing the three great marine Gulfs of Tomini, Tolo, and
Boni. Owing to this peculiar configuration, paralleled only by **its**
eastern neighbour Jilolo, no part of it is over 50 miles from the sea,
although it has a total length of about 800 miles, and an area of over
70,000 square miles, with an enormous coastline of over 2000 miles.
All the limbs terminate in islands, or insular groups, such as Salayer
in the south; Muna and Buton in the south-east; Peling, Bangay, and
further seawards the Sula Group in the east; Tagolando, Siao,
Sanguir, and others in the north-east—all evidently at one time forming part of the mainland, and indicating a former extension of Celébes
towards the Sunda, Molucca, **and** Philippine Archipelagoes.

Although the interior still awaits systematic exploration, sufficient is
known of its structure to conclude that it almost everywhere consists of
very old crystalline, carboniferous, and sedimentary rocks, except in the
extreme north-east. Here is a remarkable cluster of 11 volcanoes, several
of which have been in eruption during the present century, and one of
which, Klabat, attains an elevation of nearly 7000 feet. Elsewhere there
are several peaks, ranging from 8000 to 10,000 feet and upwards, such as
Donda (9500 ?) at the north entrance of Macassar Strait; Lompobatang
(8200), and Bonthain (apparently about 10,000), at the south end of the
southern limb; Latimojong, Tùkala, Tampoki, and Tjinrana in or near the
central nucleus. But although the country is generally mountainous, with
a mean elevation of perhaps not less than 2000 feet above sea-level, extensive level **or** slightly rolling plains occupy a large space between the

uplands and the sea. These plains are in some places covered with dense primeval forest, and elsewhere overgrown with herbs and grass, affording excellent pasture for horses and cattle. (Temminck.)

Rivers and Lakes.—Owing to its fragmentary character, Celèbes affords no space for the development of great rivers. The largest is the Sadang, which enters Mandhar Bay on the west coast, after a southerly course of 160 miles ; but the most useful for navigation is the Chinrana, accessible for good-sized native craft to the large Lake Luboya, some 20 miles from its mouth on the west side of the Gulf of Boni. Besides the Luboya, there are several other lacustrine basins of considerable size in every part of the island, Celèbes in this respect resembling Sumatra and the Philippines. Tondano in Minahassa, and Limbotto in the Gorongtalo district further west, send their overflow to the Celèbes Sea and Gulf of Tomini respectively, and most of the lakes stand at a considerable elevation in the midst of wild and romantic scenery.

Climate.—Notwithstanding its equatorial position, Celèbes enjoys a relatively cool and healthy climate, thanks partly to the high relief of the land, partly to the sea-breezes, by which the tropical heats are everywhere tempered. But for the violent earthquakes and volcanic eruptions of the northern peninsula, this island would be in every respect one of the most favoured regions in the world.

Minerals.—Iron, salt, and gold are found in abundance, the latter being widely disseminated throughout the northern districts, and more extensively exported than from any other island except Borneo. Tin and copper also occur, and mines of both are worked in several places. But although the carboniferous strata are well developed, they have hitherto yielded nothing but coal of poor quality.

Flora and Fauna.—The chief vegetable products are maize, rice, cassava, tobacco, coffee, yams, sugar, and sago, while the forests contain a great variety of valuable trees, from one of which the well-known *badean* or Macassar oil is extracted. Other useful species are the oak, teak, cedar, ebony, sandalwood, pepper, betel, areca, besides the clove and nutmeg, which grow wild, and the *upas* or "poison" tree, and bamboos in great abundance.

The Celèbes fauna differs from those of Borneo and Java in the absence of tailed monkeys, feline and canine animals, insectivora, the elephant, tapir, and rhinoceros. They are replaced by a large black tailless baboon, two kinds of cuscus (an opossum-like marsupial), two rats, five squirrels, and the already-described babirussa and sapi-utan, altogether peculiar to this island. **Of** the 160 species of land-birds, as many as 90 are also peculiar to Celébes and adjacent islands, while of the remainder 50 come from the Asiatic and 20 from the Australian regions. Three remarkable genera of starlings (*Basilornis, Enodes,* and *Scissorostrum*), two indigenous magpies (*Streptocitta* and *Charitornis*), and an anomalous kingfisher (*Cryptsis*), have no near allies in the Archipelago, and are only remotely connected with groups now inhabiting the Asiatic or African continents. The insects also differ largely from those of the Sundas and Moluccas, presenting certain peculiarities of form and colour occurring nowhere else. Thus, the more its living organisms are examined in detail, the more it becomes necessary to detach this remarkable region from the rest of the world.

Inhabitants.—Although usually classed as Malays, the bulk of the inhabitants, **both** wild and cultured, seem to belong rather to the Indonesian group. Not only the "Alfuros," a collective **term** applied by the Malays to the rude and pagan **natives**, but also **the** civilised people, such as the *Mangkassars* and *Bughis* of the southern peninsula, are described as tall, well-proportioned, **with regular** features, and comparatively fair and even white **complexion**, and altogether resembling the Samoans, Tahitians, and **other eastern** Polynesians far more than the Malays. (Dumont d'Urville.)

In general those described as Alfuros, such as the *Galelas, Tarajas, Zaili,* and many others, are heathens **at a** very low stage of culture, while the civilised communities, numerous especially in the south, profess a mild form of Mohammedanism, modified by many local rites and traditions. The Bughis especially are an intelligent, energetic, and daring race, given more to trade and seafaring than to agriculture, and renowned throughout the Archipelago for their commercial qualities, vigour, and enterprise. Long before the Mohammedan period they had attained a certain degree of culture, derived probably from the Hindus of Java, and were distinguished by their courteous habits, hospitality to strangers, and knowledge of letters. Both Mangkassars and Bughis have a peculiar writing system, somewhat resembling that of the Sumatran Rejongs, and doubtless received from the same common Indian source. Their languages belong fundamentally to the Malayo-Polynesian family, but possess many independent forms, **and** foreign or unknown elements, derived probably from an original Indonesian form of speech, diffused throughout the Archipelago before **the** arrival **of** the Malays from the Asiatic mainland.

Some of the wild tribes, especially in the central and northern districts, are head-hunters, and even cannibals, and in other respects betray a marked resemblance to the Bornean Dyaks, from whom they are probably descended.

Political Divisions.—Celébes is claimed entirely by the Dutch, and is divided by them into the Residencies of *Mangkassar* (*Macassar*) which embraces the southern peninsulas, besides Salayer, Sumbawa, and part of Floris in the Lesser Sunda group, and *Menado*, which comprises **the** northern peninsulas, with the Siao, Sanguir, and Tulur

islets between Celébes and Mindanao. A third division, extending
round the north and west sides of the Gulf of Tolo, is included in the
Residency of Ternate in the Moluccas. The chief Dutch settlements
are at *Menado* and *Gorongtalo* in the northern, and at *Mangkassar*
(*Vlaardingen*) at the southern extremity of the south-western peninsula ; but very little of the country is really occupied, or directly
administered by them, being still mostly held either by unreduced
wild tribes, or by native rajas, who, however, recognise their political
supremacy. Mangkassar is the largest town and chief seaport in
Celébes, its trading relations extending to Java, Singapore, the Aru
Islands, and New Guinea. The district produces abundance of rice,
besides some cotton, with which the native women make large
numbers of "sarongs," universally worn by Malays of both sexes.
The Dutch Governor resides at *Fort Rotterdam*. Menado, capital of
the Northern Residency, is a small place of scarcely 3000 inhabitants,
near the north-east extremity of the island. On the opposite side of
the peninsula is the station of *Kema*; and on the same side, but much
further west, the little port of Gorongtalo, almost the only settlement
in the district of like name, which is inhabited by rude tribes under
rajas supervised by a Dutch Assistant-Resident.

The south-western peninsula comprises nine petty Mohammedan States,
constituting a sort of Bughi confederacy, with capital at *Boni*, near the
head of the gulf of like name, and in alliance with the Dutch. To the
north-west is the smaller *Mandhar* confederacy of seven Mohammedan
States, comprising the western portion of the island, where it projects into
Macassar Strait, north of Mandhar Bay. The Mandhar people, who, like
their Bughi neighbours, have developed a native culture, are daring trepang
fishers, and enterprising traders.

The large islands of *Bûton* and *Môna*, forming a southern extension of
the south-eastern peninsula, constitute a Mohammedan State under a sultan
subject to the Dutch. In the same way, the eastern islands of *Peling*,
Banggai, with the *Sulla* (*Xulla*) group, belong nominally to the Sultan of
Ternate, who also recognises the supremacy of the Dutch. The inhabitants
of all these islands are Mohammedan Malays, or Indonesians, speaking
several distinct Malayan dialects.

Agriculture, Trade, Industries.—Except where Dutch influence has made itself felt, very little attention has been paid to
husbandry. The soil is much inferior to that of Java in fertility,
and the only part of Celébes yielding surplus corn for exportation is
the eastern or volcanic portion of the northern peninsula. Recently
the culture of coffee and cocoa has been introduced, but the staples of
agriculture are maize, sugar, tobacco, cotton, and especially rice, of

which three varieties are grown. The inhabitants **excel in the** manufacture of the sarong, or national garment, woven **by the women,** and together with variegated mats largely exported.

But the chief pursuit of the civilised communities is trade and maritime enterprise. The Bughis and Mangkassars are at present the most adventurous and skilful seafarers in the Eastern Archipelago With their little craft of peculiar build, and from 40 to 50 tons burden, they have acquired a large share of the local carrying trade, making long journeys as far east as New Guinea, and westwards to Sumatra, in the track of the monsoons. The outward cargoes are chiefly cotton cloths (sarongs), edible birds'-nests, trepang, coffee, rice, gold-dust, tortoise-shell, sandal-wood, matting, bees-wax, gomuti cordage, sugar, and cocoa-nut oil. In exchange for these commodities, they ship to Batavia, Singapore, and other places, cotton and silk fabrics, steel and iron ware, China goods, birds of paradise. The people of Celébes have numerous settlements in Borneo, Sumatra, **and** many other parts of the Archipelago.

MOLUCCA AND BANDA GROUPS.

Under the term MOLUCCAS (MALUCOS) or Spice Islands, are now generally comprised most of the groups lying between Celébes and New Guinea, and divided politically into the three Dutch Residencies of AMBOYNA, BANDA, and TERNATE, with a total area of over 20,000 square miles. They form two distinct geographical groups : the Moluccas proper, with Jilolo to the north ; the Bandas, with Ceram and Bûrû in the south, separated from Celébes, the former by Molucca Passage, the latter by Pitt Passage. Northwards, the Tulur islets, lying in comparatively shallow water, serve to effect a transition to the Philippines, while in the south they are severed by the deep **basin** of the Banda Sea, from Timor and Timor Laut. They are **almost** exactly bisected **by** the equator, north and south of which **they extend** in Tulur **and the** Bandas a little beyond **the** fourth **parallel of latitude.**

Physically the two large islands of Jilolo and Ceram appear **to** consist mainly of crystalline and old sedimentary rocks, while all the smaller groups are essentially igneous, forming an important section of the volcanic belt, which traverses the whole Archipelago from Sumatra to the Philippines. Many of **the** volcanoes are still active, and several were in eruption when these waters were visited by the Challenger expedition in 1874. On that occasion Ternate, a huge volcanic mass, with three superimposed cones, 5600 feet high, was ascended by Moseley and Balfour, who found that the neighbouring Tidor, one of the highest points in the whole group, attained an elevation of 5900 feet. Other conspicuous cones are Metir (2800), a little north of the equator ; Hieri (2200), north of Ternate ; and

Gûnong-Api (1860), in the Banda group, one of the most active in the whole Archipelago, and the scene of at least seven violent eruptions during the last two centuries. The outburst of 1825 was accompanied by tremendous earthquakes, which nearly destroyed Great Banda and the neighbouring islet of Neira. Many of the volcanoes are of comparatively recent date, and it seems evident that for many ages the whole region has been the scene of continuous disturbances, which have reduced it to its present fragmentary state. During Eocene, if not even Miocene times, it probably formed continuous land with Célèbes and the Philippines, the natural history of all these groups having many features in common, which are also best explained on this supposition.

Fauna and Flora.—The fauna of this region is **connected with** that of Célèbes by the babirusa, found also in Bûrû. **Other characteristic forms** are the bird of Paradise and monkeys **of Bachian,** the civet, bats, and swine, besides the marsupial cuscus **and flying** opossum. Of birds, parrots, pigeons, and kingfishers **are the prevailing** species, including the rare green-fruit dove, and **racket-tailed** kingfishers. The crimson lori, ground-thrush, fly-catcher, cassowary, and mound-builders, are also met. Here, as elsewhere **in the** Archipelago, are found butterflies of the largest size and most **vivid hues.** The beetles also are remarkable for their size and beauty : **the long-armed** beetle of Amboyna being one of the giants **of the insect world.**

This region is **the native home of the nutmeg,** clove, **and** other spices, which appear to have spread **thence to** various parts of the Archipelago and Indo-China. But in the Moluccas proper the clove is no longer produced, the plant having been extirpated by the early Dutch rulers, who desired to enhance the value of the spices by restricting their cultivation to the Banda Islands. Other valuable vegetable products are cardamoms, the kanary nut, Cayaput oil, dammer, pandani, and sago, the last-mentioned forming the staple of food in most of the islands.

Jilolo, properly **Halmahera,** largest of **the** whole group, is of **an** extremely irregular form, curiously resembling that of Célèbes, and like it developing three spacious inlets **on its** east side. The interior, which still awaits thorough exploration, is generally rugged **and** mountainous, culminating in the northern peninsula with the volcanic Gamokonora, said to have been upheaved in 1673. Jilolo, which has a total length of some 200 miles, with an extreme breadth of 90, **and** an area of about 6500 square miles, is mostly occupied by peoples **of** Malay stock, akin to those of the adjacent islands of Ternate and Tidor. But the northern parts are still held by the so-called " Alfuros," wild tribes betraying both Papúan and Indonesian affinities, and representing the aboriginal elements before the **arrival** of the intruding Malays from the west.

In the north-west lies the large island of *Mortai* (*Moro*), now separated by Mortai Strait from the northern peninsula, union with which would greatly increase the resemblance already noticed between Jilolo and Celébes. But a striking difference is presented by the aspect of the West Coasts, that of Celébes being destitute of islets, while that of Jilolo is fringed by the chain of **the**

Moluccas Proper, stretching from **the** central nucleus southwards to Batyan, which corresponds at the southern with Mortai at **the northern** extremity. Taking them in their order, from north to south, the members of this highly volcanic **and** fertile chain are *Ternate*, *Tidor*, *Makyan*, with the coralline *Kaiou*, besides some uninhabited islets and reefs. Still further **south is** the totally uninhabited *Oby* group (Great and Little Oby, &c.), midway between Sulla and Misol. The natives of the true **Moluccas and** Bachian are all Mohammedans of Malay stock, speaking **several** distinct Malay tongues, and governed by sultans under Dutch supervision. Great Oby is 45 miles long and mountainous, with peaks 5000 feet high.

The Sultans of Ternate and **Tidor** were formerly amongst the most powerful in the Archipelago, ruling over scattered territories, which comprised large tracts in east and north Celébes, Jilolo, west New Guinea, and intervening insular groups. But the Dutch are now virtually masters of both States, with a Resident at the town of Ternate, where is centred all the trade of the Moluccas in the eastern seas. This trade consists chiefly of tortoise-shell, trepang, beeswax, massoi-bark, and birds-of-paradise.

Ceram—Ké.—Next in size to Jilolo, and scarcely better known, Ceram stands in somewhat the same relation to the other islands of the southern, that Halmahera does to those of the northern group. Stretching 160 miles east and west, with an average breadth of 35 miles, and an area of nearly 7000 square miles, it **fills** up much of the space between Búrú and New Guinea, and towards the west **is nearly** divided into two unequal parts by deep inlets on the north and south coast. The surface is very mountainous, the whole island being traversed by a densely-wooded range running from east to west, at 6000 to 10,000 feet high. The sago-palm, which grows wild, supplies abundance of food for the local consumption and export, sago-cake being much used, like our "sailor's biscuits," by the native seafaring populations.

The bulk of the inhabitants are of Papúan type and speech, **with a** considerable intermixture of Malay elements, especially on the coast.

Here are a few scattered Mohammedan and Christian settlements; but the only place of any importance is *Wahai*, a fortified station on the north side over against Misol.

Ceram is continued south-eastwards, in the direction of Aru, by **a chain** of islands, **of which the** most important are *Ceram Laut*, *Goram*, *Manawolko*, *Matabello*, *Teor*, and the *Ké* group (Great and Little Ké), all mostly unsettled, and inhabited by people of mixed Papúan and Malay descent, the dark element almost everywhere predominating. In Ké there are some Mohammedan Malay, or Indonesian communities; but the majority of the people are distinctly Papúans in type and speech, and like most Papúans skilful wood-carvers. They also excel in boat-building, the materials being supplied by the forests of fine timber covering extensive tracts in this group. Ké may be regarded as the south-eastern extremity of our Oceanic division, for immediately beyond it the deep marine basin terminates at the 100-fathom-line indicating **the** north-western limits **of** the Australian world.

Búrú.—Ceram is separated on the west by Búrú Strait from the large island of Búrú, which is 85 miles by 40, with an area of nearly 2000 square miles. Although somewhat sterile, the northern districts produce the plant whence is extracted the far-famed Cayaput-oil. This part of the island is occupied by a people of Malay type, while in the south the Papúan is the dominant element. Búrú consists mainly of old sedimentary **rocks**, but touches the great volcanic belt at its western extremity, where Cape Palpettu is dominated by a lofty cone still active or quiescent. At *Cayeli*, on the north side, is a strong Dutch fort, with a Commandant under the supervision of the Resident of Amboyna. It exports considerable quantities of fish, sago, Cayaput-oil, and swine, which, being fed on sago, have a **finer** flavour than any other. The island is divided into several petty States, whose rajas spend most of their time in Cayeli, under the influence of opium.

The chief physical feature of Búrú is Lake Wakolo, a fine sheet of water, situated near the centre of the island, some 1900 feet above sea-level, and surrounded by high hills, except where it seems to escape through the Wai Nipe river. Wakolo, which was visited in 1883 by H. O. Forbes, looks like a flooded crater, several miles in diameter, and 240 to 300 feet deep. It is remarkable that no fish except eels live in its waters, which are very little navigated by the timid or superstitious natives dwelling on its shores.

Some recent ethnologists have on somewhat shadowy grounds pointed to Búrú as the cradle of the large brown Polynesian race (Samoans, Tahitians, Hawaiians, &c.), or at least the land whence these Indonesians started on their long migrations from the Archipelago eastwards to the Pacific.

Amboyna.—This historical island, where the Dutch and English long contended for supremacy in the eastern seas, lies south from the west end of Ceram, and is little over 30 miles long, by 10 or 12 broad, with an area of 210 square miles. The surface is hilly but fertile, yielding large quantities of cloves, of which a million pounds have been exported in favourable years. The clove plant, which resembles the pear-tree, grows to a height of 40 feet, bearing fruit for 100 seasons after its ninth year. Other products are cinnamon, cotton, coffee, indigo, pepper, and sago, the latter being the chief food of the Malayan and Ceramese Mohammedan natives. Amboyna, like all coraline islands in these waters, is noted for its beautiful shells, corals, and sponges, which completely carpet the bed of the sea.

The town of *Amboyna*, capital of the Dutch possessions in the Moluccas, carries on a brisk trade in cloves, cabinet wood from Ceram, and other local produce, through the Netherlands Trading Company. The adjacent islets of *Haruka*, *Sapurua*, and *Nusa Laut*, also produce cloves, and form with Amboyna the far-famed clove-gardens of the Dutch Government. Many of the native Mohammedans have become "Orang Sirani," that is, "Nazarens," or Christians, although their new religion "seems to lie on them like an awesome thraldom." (Forbes.)

Banda.—The small but **valuable Banda** group, **which** gives its name to the neighbouring waters, **was** long the exclusive nutmeg garden of the world, and here this beautiful plant still grows in the greatest perfection. The islands, which lie some 60 miles south of Ceram, are all volcanic, one of them forming the superb Gûnung Api (2000 feet), with a still active crater. The group consists altogether of twelve islets, with a collective area of not more than 18 square miles. But here is concentrated some of the most enchanting scenery in the whole Archipelago.

The produce, including sago and cocoa-nuts, besides the staple exports, nutmeg and mace, "is grown in beautiful bowers, and garnered round its umbrageous bayleted shores in long, gaudily-painted prans, which are constantly darting about, propelled by little rowers, who plunge and flash their paddles in the sun to a buoyant merry tune. The atmosphere is charged with aromatic exhalations ; its wharfs and streets are the picture of tidiness, and the very water that laps its coral shores is brighter and purer than almost anywhere else in the world." (Forbes.) But eruptions and earthquakes are frequent, and often very destructive. The chief town and centre of trade is *Nassau* in Banda Neira.

The Perkeniers, descendants of Europeans, settled in this group since 1621, enjoyed a monopoly of the nutmeg trade till 1860.

Political Divisions.—The **whole** of **this** region, officially named the Moluccas, comprises three separate Dutch Residencies as under :—

1. *Amboyna*, so named from its central island, **embraces** all the neighbouring islets, the large island of Burú, and the western portion of Ceram. It is one of the oldest Dutch settlements in the East.

2. *Banda*, includes the western half of Ceram, the Banda group proper, the Ké and Aru groups, Timor Laut, and the Serwati Archipelago, between that island and Timor.

3. *Ternate*, comprises Jilolo, with all the adjacent islands; a part of Celébes bordering on the Gulf of Tomini, with all the intervening islands; Misol, Salawati, Waijiu, and the western section of New Guinea as far as 141° E. longitude.

III. AUSTRALIAN DIVISION.

THE LESSER SUNDAS—TIMOR—TIMOR LAUT—NEW **GUINEA.**

General Survey.—This division falls naturally into two groups —a volcanic and a non-volcanic—the former washed by deep waters, and comprising all the Lesser Sundas with Timor; the latter comprising New Guinea, with Salawati, Waijiu, Misol, Aru, with Timor Laut, and washed by shallow waters. Exceptions to this symmetrical disposition are, in the first division, the Sumba group, which appear to be non-volcanic; and in the second, Timor Laut, which lies beyond the 100-fathom-line, and consequently in deep water. For reasons already stated, **Bali is** here separated altogether from the Lesser Sundas, and treated in the Asiatic division, **as** in **every** respect forming a geographical dependency of Java.

Of the two groups the non-volcanic belongs beyond all doubt physically to the Australian mainland, from which it became detached at probably a not very remote period. On the other hand, the volcanic group is connected with Australia, not physically so much as in its animal and vegetable forms. It is apparently of recent formation, upheaved through igneous agency after the subsidence of Lemuria, of which Sumba, Timor, and Timor Laut may, like Celébes, be possibly surviving fragments. Hence, in a strictly scientific grouping, these somewhat abnormal islands should perhaps be treated in our Oceanic division, although more conveniently reserved for this place.

The two groups lie altogether south of the equator, the volcanic mainly north of the 10th parallel, the non-volcanic occupying the whole space between the equator and the same parallel. They stretch for over 3000 miles west and east across 45 degrees of longitude (106—151° E.), and have a collective area of some 365,000 square miles. But this space is very unequally distributed, over seven-eighths being comprised in the non-volcanic, and less than one-eighth in the volcanic group.

THE LESSER SUNDAS, WITH TIMOR AND TIMOR LAUT.

All the members of this group, except Sumba and Timor Laut, form so many links in one of the most remarkable chains of volcanic islands on the globe, stretching from Java and Bali just under the eighth parallel of south latitude, due eastwards to the islet of Moa, 120° E. longitude, off the eastern extremity of Timor. They are separated by the narrow but deep Lombok Strait from Bali, western limit of the Asiatic world, and like that of Java, their long axis is disposed in the direction from west to east in all cases except Timor, which is only partly volcanic, and which runs south-west and north-east. Owing to the influence of the predominant wind from the arid plains of North Australia, the climate is much drier than in most other parts of the Archipelago, and in its animal and vegetable forms the whole group resembles the same region far more than any other of the surrounding lands. Here the line is drawn very sharply between Bali and Lombok, although separated from each other by a marine passage not more than 15 miles wide.

Lombok, so designated by Europeans from the name of the southern district, is usually called Selaparán by the Balinese, and by the Bughis *Tanah Sasak*, that is, " Land of the Sasaks," as its native inhabitants are called. It is of rhomboidal shape, 55 miles by 45, with an area of 2100 square miles, and a population of some 400,000, all Sasaks (Mohammedans of Malayan stock), except about 20,000 Hindu Balinese, and 5000 Malays confined to the seaports.

Recent calcareous formations prevail in the south, which is traversed west and east by a limestone range, with an extreme altitude of 1000 feet. But the north is wholly igneous, with a parallel but much loftier volcanic range, culminating in the Lombok Peak (Günung Renjani), a remarkable mountain, with four cones encircling a crater, above which rises a fifth cone, Api, continually emitting sulphurous vapours. Sangkarejan, the loftiest of these cones, is 12,460 feet high, and between it and Yayan (6500 feet) lies the upland Lake Segara, 7900 feet above sea-level. The northern and southern ranges are connected near the centre of the island by the volcanic Sessan hills, which are clothed with a dense vegetation of shrubs and grasses, and which form a waterparting, whence flow numerous unnavigable but perennial streams, contributing much to the fertility of the land.

Marking the extreme eastern limit of the Australian animal and vegetable forms, Lombok lacks the Areng palm, the lontar (*Borassus flabelliformis*), and many other characteristic Javanese plants, and is also relatively poor in orchids, ferns, and mosses. Here are no tigers or other felidæ, while the *Oriolus horsfieldi*, and other birds common to Indo-Malaya, are replaced by cockatoos, the *tripodorhynchus timoriensis*, and several species of honey-suckers, belonging to the Australian avifauna. Similar contrasts are presented by the butterflies and beetles of Lombok, as compared with the Javanese and Balinese insect world.

The natives cultivate rice, which is largely exported, besides maize, cotton, tobacco, sugar-cane, indigo, and coffee. They also rear numerous herds of cattle, buffaloes and horses, and occupy themselves with metal-work, linen-weaving, and the manufacture of bamboo and palm fibre matting. Politically the Balinese are the dominant class, the Sasaks presenting a rare instance of a Mohammedan population controlled by Hindu masters. But all alike are under the direct supervision of the Dutch ; Lombok, since 1849, forming with Bali a Residency, with seat of Government at Mataram, three miles from the west coast. The chief sea-port is the neighbouring Amplaam, which has a mixed population of Sasaks, Balinese, Bughis, and Malays. Besides rice it exports coffee, cotton, hides, and horses ; the chief imports being salt, areng sugar, arac, opium, palm oil, hardware, and European goods. The Balinese conquest dates from the first half of the 18th century, and the Brahmanical rajas were absolutely independent of the Dutch till about 1840.

Sumbawa.—Lying between Lombok and Comodo, from which it is severed by the narrow straits, Allas and Sapi, Sumbawa is 170 miles long, varying greatly in breadth from five or six to 50 miles, with a total area of 5000 square miles. Its peculiarly irregular form is due to the deep indentations on the north coast, one of which, the Bay of Bima, penetrates 15 miles inland, thus nearly severing the island into two parts. It is essentially volcanic, with several cones over 5000 feet and culminating in Tomboro, or Tambora, 9040. Tomboro, at the entrance of Dompo Bay, was the scene of a tremend-ous eruption in 1815, when most of the land was wasted, and 12,000 of the inhabitants involved in the general ruin. The climate is much drier than that of the more westerly islands, and few of the streams are perennial, but rather resemble the Arabian wadies, rushing torrents during the rainy season, waterless sandy river-beds for the rest of the year. Hence, much of the land is unproductive ; but the forests yield the valuable sapan and sandalwood, besides teak, which appears here sporadically, being absent from the islands intermediate between Sumbawa and Java.

The chief mammals are swine, deer, and a much-prized breed of ponies, exported to Java and Mauritius. The natives are mainly Malayan Moham-medans, somewhat resembling the Bughis of Celebes, but speaking several distinct languages, and grouped into four petty States (Sumbawa, Dompo, Sangar, and Bima), under the control of a Dutch Assistant-Resident, stationed at Bima, on the north-east coast. They cultivate rice and tobacco ; other articles of export being wax, birds'-nests, gold, pearls, sulphur, sapan, and sandalwood.

Floris and Comodo.—The transition from Sumbawa to Floris is effected by the little uninhabited volcanic group of Comodo, with an area of about 300 square miles, and separated from Floris by Mangerai Strait. Mangerai and Ende are alternative native names for Floris, a European term unknown to the natives. Floris, which is about 230 by 30 to 35 miles, with an area of 9200 square miles, is mainly volcanic, with two active cones and several peaks, ranging from 6000 to 10,000 feet. Copper ores appear to abound, and sulphur as well as gold also occur. The soil is fertile on the coasts, yielding good crops of rice and maize, while cinnamon, sapan, and sandalwood are amongst the most valuable forest growths. These, with beeswax and **ponies, form the** staple of the export trade.

The bulk of the inhabitants are Papuans, Floris forming the western limit of this race. On the coast are some Bughis settlements from Celébes, and the former occupation of the island by the Portugese is still attested by some half-caste Christian communities in the *Larantuka* district on the north coast. Here was the Portuguese station, and here now resides the Dutch administrator. The interior of the country is very little known; its resources remain undeveloped, and there is little local and no export trade.

Floris is continued eastwards to Timor, through a continuous chain of islets, including Solor, Adanara, Lomblem, Pantar and Ombai, which are also under the administrator of Larantuka, himself dependent on the Resident at Kupang in **Timor.**

Sumba, or **Sandalwood,** which lies some 35 miles to the south of Floris, and beyond the volcanic zone, is 130 miles by 50, with an estimated area of 5000 square miles. With SAVU, ROTTI, and SAMAO, it forms a loop-line of non-volcanic islands, sweeping round from the west end of Floris to the west end of Timor, and, like Celébes, probably representing so many fragments of a submerged Miocene continent.

Lying off the beaten track, and visited only by Bughis traders, the group is very little known; but it appears to be inhabited by a fine race of Malayan or Indonesian Mohammedans, practically independent of the Dutch, although occasionally visited by officials from Timor. They grow rice, maize, and tobacco, and have herds of buffaloes, ponies, sheep, and goats. This group also yields for exportation sandalwood, birds'-nests, beeswax, and tortoise-shell. Savu is rocky and mountainous, with an area of about 200 square miles. Samao, within three miles of Timor, is 20 miles by seven, with an area of 150 square miles, while Rotti, also close to Timor, and 60 miles by 38, has an area of over 500 square miles.

Timor.—Physically **occupies** an intermediate position between the Lesser Sundas proper **and the** "loop-line," allied to the former in its slightly volcanic, to the latter in its more highly developed sedimentary formations, different from both in the lie of its main axis, which is not west and east, but south-west and north-east. The great prevalence of old rocks, such as schists, slates, sandstones, and carboniferous limestones, combined with the fact that it lies mainly in very deep waters, seems to indicate a former connection with the vanished Lemurian continent, of which it may perhaps be regarded as the eastern limit in this direction. Timor is the largest of all the Lesser Sundas, being 300 miles by 60, with an area of over 11,000 square miles. The surface is everywhere rugged and mountainous, with numerous irregular ridges from 4000 to 8000 feet, and peaks rising considerably higher. Mount Kabalaki, in the eastern district of Manufahi, has an altitude of over 10,000 feet (Forbes), while Gúnung Allas, near the south coast, appears to be the culminating point, with an elevation of 11,500 feet.

Iron, copper, and gold occur in several places, and the uplands yield excellent wheat and potatoes. The woodlands, which nowhere develope into true forests, contain much sandalwood of fine quality, which forms, with ponies, a chief staple of export.

The substratum of the population appears to be Papúan, but intermingled in the most varied proportions with Malayan, Indonesian, and other elements. They are divided into a large number of more or less hostile tribes, speaking as many as forty distinct Papúan and Malayan languages or dialects. Some of the tribes are extremely rude, and still addicted to head-hunting, at least during war, and to other barbarous practices. In their Uma-Luli, or sacred (tabooed) enclosures, rites are performed resembling those of the South Sea Islanders.

Politically Timor belongs partly to the Dutch, and partly to the Portuguese. The western and larger section is nominally administered by a Dutch Resident stationed at Kupang, a petty little town, with a mixed Timorese, Malay, Chinese, and European population, at the western extremity of the island. It exports sandalwood, beeswax, ponies, and maize.

The eastern section is ruled by the Portuguese, whose chief settlement is at Dili, a group of hovels and wretched houses, with a ruined fort, in a fever-stricken district on the north coast. Its chief exports are wheat, potatoes, coffee of fine quality, ponies, sandalwood, and beeswax. But there are a large number of practically independent petty States ; as many as forty-seven in East Timor alone. These "Renos," or "kingdoms," are under absolute "Leoreis," or chiefs, and subdivided into Sukus, or districts, each under a Dato, dependent on the Leorei, and assisted by a Cabo and Teneste. (Forbes.)

Scattered over the Banda Sea, between Timor and Timor Laut, are several-islands and insular groups— Wetter, Roma, Moa, the Servati islets, and Babbar—which are mostly volcanic, and consequently form a natural eastern extension of the Lesser Sundas in the direction of Timor Laut.

Wetter, which is considerably larger than all the rest put together, lies

40 miles north of Timor, is 80 miles long, largely volcanic, rugged, and nearly treeless, and inhabited by a mixed Papuan race, akin to the Timorese. East of it is the lofty island of Roma, and the volcanic Moa, both occupied by Malayan peoples. These, with Babbar, lying much farther east, are sometimes regarded as outlying members of the Serwati group, which in its widest sense also includes *Kissa*, between Wetter and Moa ; *Nila* and *Serua* in the extreme north-east ; *Damma*, midway between Nila and Roma ; *Sermatta*, east of Moa. The natives are partly of Malayan or Indonesian, partly of Papuan stock, and many are nominal Christians. All these islands are now included in the Dutch Residency of Banda.

Timor Laut, or Tenimber.—Until recently the very outlines of this group were unknown. It was figured on all maps as a continuous piece of land running south-west and north-east, nearly parallel with Timor, whereas it really consists of three considerable islands ; *Yamdena* in the centre, separated by Wallace Channel from *Larat* in the north, and by Egeron Strait from *Selaru* in the south, with a cluster or chain of smaller islets on the west and north sides. Thus, the conjecture made by Captain Owen Stanley in 1841, that "when the island is properly examined, it will be found to consist of several islands, separated by narrow channels," has been fully verified by the subsequent explorations of Mr. Hartog, who first sailed through Egeron Strait in 1877, and of Mr. H. O. Forbes, who surveyed Wallace Channel and the northern districts in 1882.

Timor Laut, *i.e.* "Seaward Timor," is a low coralline group, the land seldom rising over 100 feet, except at Egeron Strait, where the cliffs are 400 feet, and at L[a]ibobar, apparently a volcanic islet on the west coast, with an extinct crater 2000 feet high. There are no streams, and the poor soil, covered with a typically coral island flora, yields little beyond maize, the staple of food, manioc, sweet potatoes, tobacco, some sugar-cane and cotton, and a little rice. The fauna includes buffaloes in a wild state, a cuscus (marsupial), some bats, the beautiful scarlet lory, here indigenous, new or rare varieties of the ground-thrush, honey-eater, and oriole. The birds seem to have come mainly from New Guinea, the insects from Timor ; a few of both from Australia.

The aborigines are evidently Papuans, with a language like that of the Ké islanders ; but there is a large intermingling of Malayan and Indonesian (Polynesian ?) elements. They are a fine, handsome people, often over six feet high, noted, like all Papuans, for their high artistic sense, betrayed especially in their wood and ivory carvings. In other respects they are pagans in a low state of culture, mostly divided into hostile communities, and addicted to piracy. There is a Dutch official ("Postholder") stationed at *Ritabel*, on the west coast of Larat, a trading station of the Bughis from Celébes.

NEW GUINEA.

General Survey.—This great island, exceeded in size by Australia alone in the eastern hemisphere, lies entirely south of the equator ; but while almost touching the line at its western, it reaches at its eastern extremity nearly to 11° South latitude. The main axis thus lies in the direction from north-west to south-east, stretching across 20 degrees of the meridian (131°—151′ E. long.) with a total length of some 1500 miles, and area roughly estimated at 325,000 square miles. Owing to its curiously irregular form, resembling in outline some extinct saurian, with head facing Jilolo and tail touching the Louisiades, the breadth varies enormously from about 20 miles at the neck to 480 at the widest part of the body. This greatest width coincides exactly with the 141° E. long. which divides the island into two nearly equal parts, and which forms the conventional line separating the Dutch, or western, from the newly-formed British and German sections. Here the central mass bulges out southwards in the direction of York Peninsula, northernmost point of Australia, from which it is separated by the shallow waters of the island-studded Torres Strait, only 80 miles wide and nowhere over 20 fathoms deep. From this central mass the head and tail project north-westwards and south-eastwards as two peninsulas, the former formed by the deep inlet of Geelvink Bay on the north coast, the latter by the broader bight of Papua Gulf on the south coast. The western peninsula is again disposed in two secondary peninsulas by McCluer Inlet running in the opposite direction from Geelvink Bay, while the eastern tapers gradually towards the Louisiades. But here also McCluer Inlet finds its counterpart in Huon Gulf indenting the coast opposite Papua Gulf. Recent exploration has also shown that the central parts of the seaboard are far less uniform than had been supposed, being diversified by numerous little bays and headlands, as well as by the mouths of many streams, whose existence had not hitherto been suspected.

Islands.—Grouped round the western extremity of New Guinea are several insular dependencies of the mainland, which they closely resemble in their physical constitution, natural history, and inhabitants. The most considerable are *Jobi*, *Biak*, *Sûk*, and *Mafor* (properly *Nufôr*) in Geelvink Bay ; *Waijiu*, *Batanta*, *Salawati*, and *Misol*, forming a westerly continuation of New Guinea in the direction of the Moluccca and Banda Archipelagoes ; lastly, the *Aru* group on the south-west coast, noted for its birds of paradise and pearl

N 2

fisheries. Here is the port of *Dobbo*, "the Nishni Novgorod of Malaysia" (Guillemard), much frequented by Bughis and Chinese dealers. The large island of *Frederick Henry* on the south coast almost forms part of the mainland ; and Torres Strait, further east, is thickly strewn with numerous islets, partly coralline, partly of old formation, probably fragments of the miocene land which at this point formerly connected New Guinea with Australia. Of these the largest are *Thursday, Banks, York, Darnley,* and *Murray,* which are all politically attached to the government of Queensland.

The coralline islets of Torres Strait are often wrongly described as a western continuation of the Great Barrier Reef of East Australia. Between these groups there flows deep water mostly free of islands, while the sunken Barrier Reef of south-east New Guinea, about 140 miles in length, reaches no further west than Cape Possession near Hall Sound, 146° 20′ E. Here it is arrested by the copious fresh-water streams, which discharge into Papua Gulf, and destroy the work of the coral zoophytes. The north coast of New Guinea east of Geelvink Bay is almost destitute of reefs and islands ; but numerous groups, such as the *D'Entrecasteaux* and *Louisiade* Archipelagoes, are clustered round the south-eastern extremity of the mainland. Off the north-east coast are the large islands of *New Britain* and *New Ireland* now occupied by the Germans, and by them re-named the "*Bismark Archipelago.*" But all these groups belong rather to the Pacific insular world, and are most commonly included in the *Melanesian Division* of the South Sea Islands.

Physical Features.—Till recently New Guinea was a *terra incognita* in the strict sense of the term, and even still by far the greater part of the interior remains to be explored. Hence any attempt at a detailed account of its relief would be premature. It is known, however, to be essentially a highland, partly even an Alpine region, developing plains, or low-lying tracts, chiefly along the lower courses of the rivers, and elsewhere traversed by lofty, and in some places snowy, ranges running mostly north-west and south-east in the line of the main insular axis. These ranges appear to form more or less continuous single chains in the north-west and south-east, while in the central region they diverge into parallel systems, at some points approaching close to the seaboard and enclosing extensive plateaux and even low-lying level tracts. The best known sections are the Arfak hills (9000 to 10,000 feet) back of Geelvink Bay in the north-west, and in the south-east the Sir Arthur Gordon, Albert, Yule, and Owen Stanley ranges, the latter culminating with the double-crested Mount Owen Stanley (13,205 feet), approached, but not yet ascended, by Chalmers and Forbes. In the vast unvisited **central region other** great ranges, such as the

Charles Louis (17,000 to 18,000 feet), **are** traced on the maps, and described as towering above the line of perpetual snow by geographers relying on the somewhat indistinct reports of travellers. At the same time the existence of such Alpine heights is rendered highly probable by the presence of copious perennial streams flowing in independent channels to the coast, and which are found to be **far** more numerous than had till lately been supposed.

Rivers.—The largest river in New Guinea appears to be **the Fly**, which enters the west side of Papua Gulf through a large **and intricate** delta, and which D'Albertis ascended in 1876 for 500 miles in a steam-launch. It drains a vast **swampy region** diversified with wooded mountains and treeless plains **broken** by isolated hills, the scenery in many **places** presenting **an** Australian aspect. Another large river, the Empress Augusta, was discovered so recently as 1866, on the north-east coast, by Dr. Finsch, and navigated for 40 miles by Captain Dallman, who reported it navigable for a much longer distance. Mr. Morris, Dutch Resident of Ternate, also surveyed in 1883-84, several hitherto unknown rivers on the north coast, such **as the** Wiriwaai and Witriwaai, apparently two branches of the same stream, and the much larger Aiberan (Amberno or Mamberan), that **is** "Great River," which he ascended for 60 miles, and found to be 800 yards wide and seven fathoms deep near its mouth. Two **large** rivers, the Davadava and Hadava, not marked on any map, **also** reach the sea at Milne Bay, the latter with an intricate delta 12 to 16 feet deep, and apparently leading into the heart of the country. But owing to the action of the south-west monsoons the mouths of the coast streams are mostly silted up with sand and mud, hence unnavigable. Altogether it may be anticipated that the more the interior is opened up the more it will be found covered with " mountains, north, east, south, and west " (Chalmers), and traversed by copious perennial streams flowing from the central water-parting to the northern and southern seaboards.

Geological Formations.—The salient formations appear to be a substratum of granite and gneiss cropping out in the Arfak hills **and** elsewhere ; stratified clay slates, and both old and recent limestones and calcareous Lower Miocene clays with fossil shells identical with those of south-east Australia. Quartz, greenstone, and jasperoids also occur on **the** south-east coast, resembling those of the Silurian and Devonian series of the New South Wales gold-fields.

Gold will probably be found both here and in the Hadava river-basin as well as on the uplands and north-east coast. It is usually asserted that no active volcanoes, or even any external cones, occur in New Guinea ; but this is a mistake due to hasty generalization from imperfect surveys, for the spurs projecting on either side from Mount Owen Stanley contain several craters said to be formed by recent volcanic action. Pumice, also, and other igneous matter, cover the slopes of the Finisterre hills, while earthquakes are of frequent occurrence in many places.

Climate.—On all the low-lying coast-lands and about the river-mouths the climate is malarious, and unsuitable for European settlers. On the uplands the tropical heats are tempered by the **marine** breezes, which in the northern and western districts accompany the north-west, and in the southern and eastern the south-east, monsoon. The latter prevails from July to September, **and is** often very violent, arresting all navigation in Torres Strait.

The heats are rendered more oppressive by the heavy rainfall, and Guillemard, who lately visited the north coast, found the climate more trying than that of any other region except the Persian Gulf **in** summer. " Bathed in perspiration from morning till night and from night till morning, we woke utterly unrefreshed by sleep. The temperature, which in a dry climate would not have been unpleasant—for it was rarely above 90° F.—was intolerable. Everything to which damp could cling became mouldy, and our boots, if put on one side for a day or two, grew a crop of mildew nearly half an inch in thickness " (ii. p. 291).

Flora.—The original vegetation appears to have been mainly Malayan, which still largely prevails in most districts. But numerous Polynesian, Asiatic, and Australian species have also invaded the island, and all these different floras are found in some places intermingled. Thus W. Wyatt Gill speaks of taro, yams, gigantic aroids, the ivory nut-palm, cotton, tobacco, the oak tree, capsicums, strawberries, raspberries, and the nutmeg, all occurring in and about the Laroki valley near Port Moresby on the south-east coast. Elsewhere on the same coast, J. Chalmers met during a single stroll, " a strange profusion of coconut, sago, and betel palms, numerous bread-fruit, and large *tamanu* trees, *dracœna*, and crotons of **various kinds,** ferns in abundance, and mangroves." The Australian eucalypti and acacias and the Oceanic coconut are everywhere familiar sights along the south coast, and the immense variety of vegetation is further attested by the presence of the pandanus with its strange aerial roots ; the costly red cedar (*Cedrela Australis*) ; the *potipoti*, growing to a height of 60 feet and yielding a much-prized fruit ; the *Cordyline terminalis*, jack fruit, and banana all widely diffused ; the **zamia,** forming a curious link between palms and ferns ; the *Erythrina*, *Barringtonia speciosa*, and other flowering forest trees ; **lastly,**

the native jute plant, with edible root, and stalk yielding the finest jute fibre in the world (Gill).

The subalpine flora is represented by oats, rhododendrons, araucarias, umbelliferæ, &c., while the chief cultivated plants are maize, millet, sugar cane, taro, rice, pumpkins, yams, and the sago-like *sali*. The sago palm itself, although yielding a staple of food, appears not to be cultivated, but to grow wild in the marshy districts. The natives till the land with great skill and neatness; but the few patches thus reclaimed in the more favoured localities are of no account compared with the rest of the land which is still mostly overgrown with dense primeval forests diversified in some places with grassy and treeless tracts of limited extent.

Fauna.—While the flora is to a large extent Malayan, the fauna is in some respects essentially Australian, the older marsupials still everywhere holding their ground against the higher mammals, which appear to be represented almost exclusively by the pig, the dingo, mice, the flying-fox, and other members of the bat family. There are at least three species of cuscus, two of the wallaby, and several varieties of the true kangaroo and other marsupials, besides three species of the spiny ant-eater, allied to the Australian Echidna, which, like the Platypus, are now known to be oviparous, thus supplying a further link between reptiles and mammalians (W. H. Caldwell). Of true reptiles by far the largest and most formidable is the crocodile, which infests nearly all the rivers, attaining a length of over nine feet, and both devouring and is devoured by the natives. Snakes, which occur in great variety, are also eaten, and even by the cannibals preferred to pig or any other except human flesh.

The avifauna, which is specially rich and beautiful, presents nearly 500 indigenous species, mostly belonging to Australian genera, besides many locally-developed varieties. Malayan forms also occur, together with others common to the whole Oceanic domain. But the special glory of this avifauna are the birds of Paradise, of which there are at least twenty species, all restricted to New Guinea and its islands, with the single exception of the standard-wing found in Jilolo and Bachian. Other more or less characteristic forms are the cockatoos, parrots, lories, the spur-winged plover, kingfishers, mound-builders, honeysuckers, flycatchers, crested and other pigeons, comprising altogether about forty genera of exclusively Papuan land-birds (Wallace). The gorgeous plumage of the feathered tribe is rivalled by the resplendent colours and metallic lustre of the numerous local varieties of butterflies and beetles. A curiosity of the shell world is the kima, a gigantic clam, often measuring 32 inches by 19 (Gill).

Inhabitants.—The great bulk of the natives belong undoubtedly to the **Papuan** stock; but such are the discrepancies presented by the different tribes **in their** physical appearance, mental qualities, and grades of culture, that D'Albertis and some other observers have felt inclined to doubt the existence of a Papuan type at all. These points, together with a general account of the Papuan populations, have been discussed at page 123, and need not be further dwelt upon in this place. As regards New Guinea more particularly, it will suffice to observe that the numerous and often profound departures from the normal Papuan standard may be attributed mainly to long isolation **in** separate tribal groups, and to constant crossings with other peoples, such as the Karons and other Negritoes in the interior, Malays and "Alfuros" along the western seaboard, and Indonesians (brown Polynesians) especially on the south-east coast. But **not**withstanding these diverse interminglings there is a general prevalence of the more salient Papuan characteristics—mop-head, arched nose, long and high skull, sooty-black complexion—from Misol **and** Aru in the extreme **west** to **the** Louisiade Islands in the extreme **east.** The same features **are** found diffused throughout **Melanesia in the** Pacific, and as far west as Floris in the Eastern Archipelago.   Hence New Guinea has been regarded as the natural, as it certainly is the geographical, centre of the Papuasian world. But from this it does not follow that here the type first became specialized, and there is even reason to suppose that the earliest inhabitants of New Guinea were not Papuans but Negritoes. This, however, is a point that can be determined only by further exploration in the interior, where some Negritoes have already been found (Dr. Hamy).   The general movement of Papuan migration may, consequently, not have been from New Guinea west and east ; but either from Melanesia westwards, or from the eastern Archipelago eastwards. **And so far as** New Guinea **is** concerned this diffusion of the race **must be referred to a** period posterior to the separation from **Australia, for** the indigenous populations of these two regions belong **to totally** different branches of the Negro family. The transition **from the** true Papuans of Torres Strait to the true Australians of the **main**land **is** extremely abrupt, and for this and other reasons it seems evident that the two great islands were peopled by independent waves of migration at some time subsequently to the subsidence of the land now flooded by Torres Strait and the Arafura Sea. Probably **both** were uninhabited till very late tertiary or early quaternary times.

The New Guinea natives have been hitherto carefully studied only at a few points on the seaboard, such as round the shores of Geelvink Bay, at

Humboldt Bay, and especially along the south-east coast. The result is on the whole unfavourable, their general social condition appearing to be much lower than had been supposed. Some of the practices associated with their treatment of the dead, as alluded to by the English missionaries about Port Moresby and Redscar Bay, are indescribably revolting, and seem to place these savages at the very lowest stage of human culture. Apart from provocation from Europeans, they are also found to be naturally false and treacherous, of filthy habits, and unclean eaters, devouring vermin and all things digestible, while giving a decided preference to reptiles, pig, and man. In some places cannibalism in its most repulsive form is universally practised, and to attend one of their periodical cannibal feasts an invitation was sent to the Rev. James Chalmers, who found the guests strutting about " with pieces of human flesh dangling from their neck and arms." A child destined for this banquet "was spared for a future time, it being considered too small." (*Work and Adventure in New Guinea*, 1885.) Needless to say that amongst these communities Christianity has not made much progress. Some of the north-eastern tribes are so backward that they use nothing but shell implements; they could hardly be made to understand the purpose of a tomahawk, and were scared by a match being struck by a member of Captain Bridge's surveying party (1884). Yet of good augury for the future is the fact that both the true Papuans and the half-caste Polynesians manufacture some articles, and especially pottery, not only for local use, but for the express purpose of trading with their neighbours.

Political Divisions.—While most of the country remains in the hands of the natives, the whole island has since 1885 been nominally distributed amongst three European powers. The claims of the Dutch to the western half, as far east as 141° E., long., claims based on the former rights or pretensions of the Sultan of Tidor, are now fully recognized. The eastern half is divided in equal proportions between England and Germany, a conventional line drawn from the Dutch frontier eastwards forming the boundary between the British protectorate on the south-east and the German on the north-east coast. New Guinea is thus parcelled out in the following proportions amongst these three States :—

					Sq. Miles.
Dutch **New Guinea**	.	.	.	.	148,000
British "	.	.	.	.	88,500
German "	.	.	.	.	88,000
			Total.		324,500

There is no Dutch settlement in New Guinea, *Dorey* at the north-west entrance of Geelvink Bay being only a missionary station, noted in the records of local exploration as the starting-point of many expeditions to the interior. The German New Guinea Company has already founded three small settlements, at *Finsch Harbour, Hatzfeldt Harbour* (4° 21′ S., 145° 9′ E.), and *Constantine Harbour* (5° 30′ S., 145° 45′ E.), while the British posts at *Yule Island, Port Moresby,* and *Redscar Bay* continue to be chiefly centres of missionary enterprise. The varied mineral and vegetable resources of the country must remain undeveloped pending the construction of roads along the coast and to the interior. From the Report of the late Sir Peter Scratchley, first Special Commissioner to British New Guinea, it appears that, owing to the unfavourable climate, the development of these resources will even then have to depend almost exclusively on coloured labour. No fixed scheme of administration has yet been adopted, the settlement of this question depending on negotiations now in progress between the Home Government and the Australian Colonies. Meantime an attempt will be made to govern as far as possible through the native chiefs, of whom there are three classes, those enjoying a purely personal, a social, or a religious influence, these qualifications being occasionally vested in the same person. (*Seymour Forbes' Report on British New Guinea*, 1886.)

Historical Note.—New Guinea was probably first sighted by A. Dahren in 1511, and first visited by the Portuguese Don Jorge de Meneses (1526?) and the Spaniard Alvaro de Saavedra (1528), receiving its name in 1546 from Ortiz de Retez (Roda), either from the appearance of its negroid inhabitants, or from a fancied resemblance of the northern seaboard to that of Upper Guinea on the West Coast of Africa. It was "annexed" by two commanders in the East India Company's service in 1793, when the island of Manasoari in Geelvink Bay was occupied for some months by British troops. But in 1814 the English Government admitted the Dutch claims to the *Raja Ampat*, or "Four Kingships" of Waijiu, Salawati, Misol, and Waigamma, including certain tracts on the mainland. As suzerain of the Sultan of Tidor, the Dutch also claim the western half of the island, to the remaining portion of which British and German protection were extended in the year 1884.

STATISTICS OF THE EASTERN ARCHIPELAGO.

AREAS AND POPULATIONS.

	Area in Sq. Miles.	Pop. (1880.)
Borneo . . .	290,000 (?)	2,000,000 (?)
Sumatra with dependencies	150,000	2,746,000
Java with Madura .	54,000	18,867,000
Banka . . .	5,200	62,000
Biliton . . .	2,600	25,000
Lesser Sunda Group .	40,000	2,200,000
Celèbes and dependencies	75,000	500,000
Molucca and Banda Groups	25,000 (?)	400,000
Philippines with Sulu .	116,000	6,100,000
New Guinea with dependencies }	325,000	500,000 (?)
Total Eastern Archipelago	1,082,800	33,460,000

DUTCH POSSESSIONS.

JAVA.

Residencies.	Area in Sq. Miles.	Pop. (1880.)
Bantam . . .	3,300	750,000
Batavia . . .	2,600	962,000
Kravang . . .	2,000	280,000
Cheribon . . .	2,700	1,240,000
Preanger . . .	8,500	1,242,000
Tagal . . .	1,500	912,000
Pekalongan . .	700	510,000
Samarang . . .	2,000	1,270,000
Japara . . .	1,200	832,000
Banjumas . . .	2,200	992,000
Bagelen . . .	1,300	1,190,000
Kedu . . .	800	695,000
Jokjokarta . .	1,200	450,000
Surakarta . .	2,500	920,000
Rembang . . .	3,000	1,010,000
Surabaya . . .	2,250	1,620,000
Madiûn . . .	2,600	955,000
Kediri . . .	2,700	700,000
Panurûan . . .	2,300	652,000
Probolingo . .	1,160	435,000
Besûki . . .	1,600	410,000
Banjûwanghi . .	1,800	60,000
MADURA . . .	2,100	770,000
Java and Madura .	54,000	18,867,000

SUMATRA.

Residencies.	Area in Sq. Miles.	Pop. (1880.)
West Coast . .	46,200	1,116,000
East Coast . .	16,280	162,000
Benćulen . . .	9,580	148,000
Lampûngs . .	9,980	120,000
Palembang . .	61,150	620,000
Achin . .	6,370	580,000
Sumatra with dependencies.	**149,560**	**2,746,000**

Residencies.		
Rian-Lingga . .	17,330	93,000
Banka and Biliton .	7,800	87,000
Borneo, West Coast .	58,900	378,000
Borneo, South and East	145,000	610,000

CELÈBES, SUMBAWA, AND BÚTON.

Residencies.		
Mangkassar . .	45,150	390,000
Menado . . .	26,600	550,000

MOLUCCAS, BANDA, AND WEST CELÈBES.

Residencies.		
Amboyna ⎫ Banda ⎬ . . Ternate ⎭	42,500	350,000
Timor (part of) ⎫ Sumba ⎬ . Savu, Rotti ⎭	22,000	250,000
Bali ⎫ Lombok ⎬ . .	4,000	1,360,000
Timor Laut, ⎫ Aru, and Ké ⎬ . .	5,500	60,000
West New Guinea .	148,000	200,000 (?)
Total Dutch Possessions	**727 340**	**25,941,000**

Total population (1884), estimated 27,500,000.

JAVA.

Chief Towns.	Pop. (1880).	Progress of Population (Java).	
Surakarta	124,000	1853	10,290,000
Surabaya	122,000	1861	13,000,000
Batavia	97,000	1871	16,452,000
Meester Cornelis	70,000	1875	18,334,000
Samarang	68,000	1880	18,867,000
Jokjokarta	45,000	1884 (est.)	20,931,000
Passurûan	39,000		
Pekalongan	31,000		
Tuban	21,000		
Bangkalan	20,000		

SUMATRA.		CELÊBES.	
	Pop. (1880).		Pop. (1880).
Palembang	30,000	Mangkassar	20,000
Achin	20,000 (?)	Menado	2,500
Padang	12,000	Amboyna	13,000
Bencûlen	6,000	Ternate	9,000
		Kupang	7,000

Population of Java, according to races (1884).

Javanese ⎫	
Sundanese ⎬	20,931,000
Madurese ⎭	
Chinese	214,470
Europeans	37,680
Arabs and sundries	14,000

Average net Revenue, £850,000.
Yearly exports (Java), £10,000,000 to £12,000,000.
Exports to Great Britain (1884), £3,184,000.
Imports from Great Britain, £2,097,000.
Nutmegs exported from Banda (1884), £76,260.
Shipping (1883), 4158 vessels of 2,740,000 tons cleared.
Railways (Java, 1884), 560 miles.
Telegraph Lines (1883), 5760 miles ; messages, 383,500.
Post Offices, 226 ; letters carried (1883), 4,729,650.
Army, 27,000, of whom 11,000 Europeans.

GERMAN POSSESSIONS.

	Area in Sq. Miles.	Population.
North-east New Guinea	88,000	100,000 (?)

PORTUGUESE POSSESSIONS.

	Area in Sq. Miles.	Population.
East Timor	6,300	300,000

SPANISH POSSESSIONS.

Districts.	Area in Sq. Miles.	Pop. 1882.
PHILIPPINES :		
Luzon	51,500	3,475,000
Visayas	23,500	2,049,000
Mindanao	41,000	154,000
Adjacent Islands	1,500	21,000
Calamianes and Palawan	5,650	61,000
Sulu Islands	980	102,000
Total	124,130	6,300,000

Manila, Pop. (1880), 270,000.
Total exports to Great Britain (1884), **£1,143,000.**
„ imports from „ „ **£1,120,000.**
Telegraph lines (1884), 720 miles.

BRITISH POSSESSIONS.

	Area in Sq. Miles.	Pop. 1881.
Saráwak	40,000	300,000
North Borneo	26,000	200,000
Labuan	30	6,000
South-east N. Guinea	88,500	150,000 (?)
Total	154,530	656,000

Saráwak	Income	(1884)	£55,000
„	Expenditure	„	58,000
„	Imports	„	344,000
„	Exports	„	325,000
North Borneo	Income	(1886)	26,000
„	Expenditure	„	43,000
„	Imports	„	120,000
„	Exports	„	80,000
Labuan	Income	(1884)	4,600
„	Expenditure	(1885)	4,200
„	Imports	(1684)	85,000
„	Exports	„	86,000

LONDON : PRINTED BY EDWARD STANFORD, 55, CHARING CROSS, S.W

STANFORD'S SCHOOL MAPS.

STANFORD'S "LARGE" SERIES.

Size 50 Inches by 58.

Price, Mounted on Rollers and Varnished, 13s. each.

The World, Mercator.	Scotland.	South America.
Eastern Hemisphere.	Ireland.	Australasia.
Western Hemisphere.	Asia.	Victoria (Australia).
Europe.	Holy Land.	New South Wales.
British Isles.	India.	New Zealand.
England and Wales.	Africa.	British Possessions, on a
London.	North America.	uniform scale.

STANFORD'S "INTERMEDIATE" SERIES.

Size 34 Inches by 42.

Price, Mounted on Rollers and Varnished, 9s. each.

The British Isles.	Old Testament (Palestine).	Acts and Epistles.
England and Wales.	New Testament (Palestine).	Australia.
Scotland.	Journeyings of the Children	New Zealand.
Ireland.	of Israel.	

Also a Map of the World on Mercator's Projection, size 50 inches by 32.
Price, Mounted on Rollers and Varnished, 12s.

STANFORD'S "SMALLER" SERIES.

Size 27 Inches by 32.

Price, Mounted on Rollers and Varnished, 6s. each. In sheet, coloured, 2s. 6d. each.

Eastern Hemisphere.	Ireland.	Acts and Epistles.
Western Hemisphere.	Asia.	India.
Europe.	Holy Land.	Africa.
British Isles.	Old Testament (Palestine).	North America.
England and Wales.	New Testament (Palestine).	South America.
London.	Journeyings of the Children	Australia.
Scotland.	of Israel.	New Zealand.

The Hemispheres can also be had mounted as one Map, size 54 inches
by 32. Price, Coloured on Roller, Varnished, 12s.

STANFORD'S "PHYSICAL" SERIES.

Size 50 Inches by 58.

Price, Mounted on Rollers and Varnished, 30s.

British Isles.	Europe.		North America.
England and Wales.	Asia.	Africa.	South America.

Also, size 34 Inches by 32.

Price, Mounted on Rollers and Varnished, 18s.

Scotland.	Ireland.

THE "EXTRA LARGE" SERIES.

THE BRITISH ISLES. Scale, 8 miles to an inch; size, 75 inches by 90. Price, Mounted
on Rollers, Varnished, 42s.
ENGLAND AND WALES. Scale, 3¼ miles to an inch; size, 8 feet by 9 feet 6 inches.
Price, Mounted on Rollers, Varnished, 63s.
ENGLAND AND WALES. Scale, 5 miles to an inch; size, 6 feet 10 inches by 8 feet.
Price, Mounted on Rollers, Varnished, 42s.
BIBLE LANDS. Size, 7 feet by 7 feet. Price, Mounted on Rollers, Varnished, 28s.
THE WORLD, IN HEMISPHERES. Size, 8 feet 4 inches by 4 feet 10 inches. Price,
Mounted on Rollers, Varnished, 20s.

LONDON: EDWARD STANFORD, 55, CHARING CROSS, S.W.

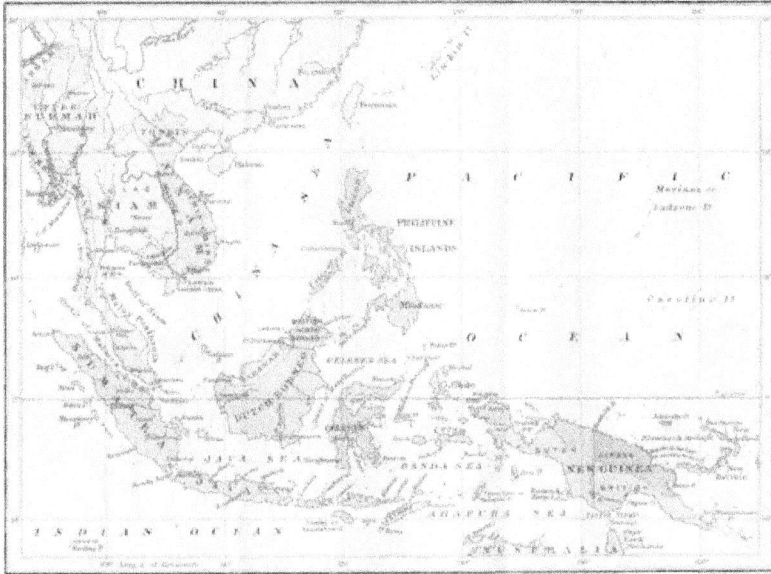

London: Edward Stanford, 55, Charing Cross, S.W.

www.ingramcontent.com/pod-product-compliance
Lightning Source LLC
Chambersburg PA
CBHW021707210326

41599CB00013B/1555